biodata mining
and visualization
novel approaches

SCIENCE, ENGINEERING, AND BIOLOGY INFORMATICS

Series Editor: Jason T. L. Wang
(New Jersey Institute of Technology, USA)

---

biodata mining and visualization

novel approaches

ilkka havukkala
Intellectual Property Office of New Zealand, New Zealand

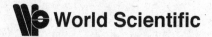 **World Scientific**

NEW JERSEY · LONDON · SINGAPORE · BEIJING · SHANGHAI · HONG KONG · TAIPEI · CHENNAI

Published by

World Scientific Publishing Co. Pte. Ltd.

5 Toh Tuck Link, Singapore 596224

USA office: 27 Warren Street, Suite 401-402, Hackensack, NJ 07601

UK office: 57 Shelton Street, Covent Garden, London WC2H 9HE

British Library Cataloguing-in-Publication Data
A catalogue record for this book is available from the British Library.

Science, Engineering, and Biology Informatics — Vol. 5
BIODATA MINING AND VISUALIZATION
Novel Approaches

Copyright © 2010 by World Scientific Publishing Co. Pte. Ltd.

ISBN-13 978-981-279-036-1
ISBN-10 981-279-036-5

Printed in Singapore.

Preface

This book is aimed at both data miners/complex systems analysts wanting to analyse postgenomic biodata and at computational biologists interested in new complex data analysis methods from other disciplines. Novel and innovative data mining and visualization techniques from artificial intelligence and knowledge engineering are covered.

In the first two chapters, the complexity of biodata and the available large and exciting datasets are reviewed, followed by local pattern analysis methods for short DNA, RNA and protein sequences. The fourth chapter explores large-scale genome comparison methods and the fifth chapter covers small scale molecular structure searches. The next chapter discusses large-scale metadata analysis and ontology-based searching and classification, including latent semantic indexing. The seventh chapter explores genomic data analysis by Support Vector Machines, microarray gene selection methods and analysis of genomes as long time series, among other techniques. The penultimate chapter deals with the integration of multimodal systems biology data. The ninth and last chapter then assesses the future challenges and includes an extensive up-to-date review on biocomputing using Graphical Processing Units and virtual machines.

Strengths and weaknesses of various approaches are reviewed, promising research trends are identified and copious up-to-date references are given for further reading. After perusal of the book,

the reader can select the most promising algorithms and datasets for their interests and is prepared to tackle the future challenges of bioinformatics in the postgenomics era.

Ilkka Havukkala
May 2010

Acknowledgements

This book is dedicated to my late parents, who always encouraged and supported me in my expatriate research career in various fields and countries. I am grateful to all my past and present scientific working colleagues in Finland, UK, USA, Japan and New Zealand, as well as many other colleagues and collaborators around the world for interesting insights and inspiration. The various journal publishers are acknowledged for their permissions to reproduce illustrations in this book. The book cover design by Azouri Bongso is acknowledged. The greatest thanks must go to my dear wife Susan for her tireless love and care since the beginning of my work in bioinformatics in Japan.

About the Author

Mr. Ilkka Havukkala is Finnish by origin, studied at Helsinki University and received his MSc and Licentiate degree in Biological Sciences, obtained a PhD in Applied Zoology from University of London in 1981 and became a docent of Zoology at Helsinki University in 1984. After research on pest insect behaviour and biological control in Finland at Agricultural Research Centre and an adjunct visiting professorship at Michigan State University, he moved to Japan in 1988 to work on insect pathogenic fungi and molecular biology of chitinolytic enzymes. In 1991 he joined the Japanese Rice Genome Program in Tsukuba Science City to start his bioinformatics research in plant genomics. In 1996 he moved to New Zealand to a biotechnology company to become the head of bioinformatics in large-scale DNA sequencing and genome analysis projects for agroforestry and dairy industries. He has been European Union Advising Expert on epigenomics and medical informatics in 2004–2005. From 2005 he was Associate Director of Bioinformatics at the Knowledge Engineering Research Institute at Auckland University of Technology, familiarizing himself with artificial intelligence methods for biodata analysis and from 2007 has been Senior Patent Examiner for biotechnology patents in the Intellectual Property Office of New Zealand (IPONZ). His interdisciplinary expertise includes plant, animal and medical genomics and biotechnology, bioinformatics, ontologies and data mining as well as intellectual property. He has published widely on a variety of topics in entomology, genomics, biotechnology and bioinformatics, is an inventor in numerous genomics patents and reviews regularly for genomics and data mining journals and conferences. His current interests include Semantic Web technologies for bioinformatics and intellectual property as well as artificial intelligence methods for integrated postgenomic biodata analysis.

Contents

Contents

Chapter 1

Introduction to Modern Molecular Biology

1.1 Cells store large amounts of information in DNA

Molecular biology and bioinformatics have become the foundation of modern biology, as scientists have started to unravel the dynamic processes of molecular scale interactions in the marvelously complex cells, the basic units of all life. In essence, biological organisms package enormous amounts of information in a chemical form into microscopic packages, the cells. Cells are bounded by a double-layered lipid membrane which encloses the cytoplasm containing the biochemical machinery for cell maintenance, growth and duplication.

Cellular life depends crucially on proteins, which serve as structural components, as enzymes for chemical reactions in metabolism and as signalling messengers relaying information within and between cells. Thus cells can be said to be protein-based machines, which obtain energy from light or chemical compounds and use it to grow and multiply. The protein machinery is controlled by the macromolecules DNA and RNA which are the repositories of the large amount of genetic information, as explained below.

The main control center of the complex cellular biochemical machine is the genetic material of DNA (Deoxyribo Nucleic Acid), which contains the information about all the proteins to be produced and about the control systems in the metabolism of the cell. The information coded in the totality of DNA in one cell (called the genome) determines the identity and characteristics of cells, and is responsible for transmitting this information to the next generation of cells. Thus the genome functions to preserve the identity and special characteristics of the self-replicating cell. An additional important property of the DNA is its mutability. Changes in DNA modify the genome information slightly and cause the next generation of cells to be slightly different. These variant cells are then exposed to natural selection to screen out the most successful cells based on the best changes in the genome. This natural selection of cells with the best genomes in each generation allows cells and organisms to adapt their cell machinery to new situations in successive generations. This in essence makes evolution possible.

The other main information-processing components of cells are RNAs (Ribo Nucleic Acid) which act as snippets of information delivered from DNA to cell protein machinery. They also have important regulatory and enzymatic activities by themselves, and represent probably the most ancient cell regulatory molecules.

In addition to proteins and DNA and RNA, cells contain a large variety of other kinds of polymers and lipids, which serve as structural components, *e.g.* supporting scaffold structures and enclosing membranes of various cell organelles. Cells also include many different kinds of organic chemicals, both simple and complex, which are processed and utilized in the cell metabolism.

The total chemical complexity in cells is huge: a single cell of a higher eukaryotic organism can contain 30,000 genes in the DNA, corresponding to at least 100,000 different RNAs, about 200,000 protein variants (including splice variants with differing amino acid sequences and variants with post-translational modifications, like phosphorylation, glycosylation etc., as explained below) and at least some 20,000 other organic compounds used in the metabolism. This is discussed further in the next subchapter 1.2.

The three main branches in the taxonomy of cellular organisms consist of single-celled bacteria and archaebacteria, and eukaryotes, most of which are multicellular. Bacteria and archaebacteria are characterized by a single large DNA molecule as the main chromosome within the cell cytoplasm, sometimes with smaller DNA plasmids containing additional genetic information. Higher organisms (eukaryotes) include fungi, plants and animals, and also a large number of relatively little studied unicellular microbes. Eukaryotes are characterized by a membrane-bound separate compartment, the nucleus, which serves as a storage site for the DNA which is packaged in several chromosomes. Eukaryotic chromosomes are much more complicated than in prokaryotes, because they have a compact wound structure with many associated proteins called histones. This is to package the information more efficiently and to keep some of the information locked safely away when it is not needed.

In addition to cellular organisms, there are many kinds of viruses, which are basically parasites of cells. They consist only of DNA or RNA as genetic information and some packaging proteins protecting the genetic material. Viruses are very small, lack cytoplasm and enzymes and therefore do not have their own independent metabolism. They can be considered to be "alive" (metabolically active) only when they are in contact with their cellular hosts.

Thus DNA in prokaryotic and eukaryotic cells is essentially a system for information storage and propagation. DNA is a long macromolecule having a sequence of different nucleotides coding information using four different nucleotides: adenine (A), thymine (T), guanine (G) and cytosine (C). The DNA molecules can be very long, up to hundreds of millions nucleotides, depending on the complexity of the organism. The typical ranges of information content in genomes of different organisms are given in Table 1-1.

Bacterial chromosomes normally have about 1–2 million nucleotides, sufficient for encoding all the instructions for microbial life. Scientists aiming to build synthetic microbes believe that only about 113,000 nucleotides, coding for less than a thousand genes, are necessary to control a simple self-replicating organism (Forster & Church 2006). In single-celled simple organisms like bacteria chromosomes are located in the main

Table 1-1. Typical information storage capacity in different organisms in counts of nucleotides. Data from various sources (*e.g.* Plant DNA C-values Database and Animal Genome Size Database, Zonneveld *et al.* 2005, Gregory 2006, Zubáčováa *et al.* 2008; minimal synthetic microbe size from Forster & Church 2006).

Taxon	Small	Typical	Big
Viruses	5,000	50,000	200,000
Bacteria	500,000	1.5×10^6	6×10^6
Archaebacteria	500,000	1×10^6	6×10^6
Minimal synthetic microbe	NA	113,000	NA
Eukaryotes			
Protista	2.2×10^6	?	177×10^6
Algae	9×10^6	850×10^6	19×10^9
Plants	120×10^6	5×10^9	250×10^9
Fungi	12×10^6	?	65×10^6
Invertebrates	29×10^6	?	37×10^9
Vertebrates	342×10^6	1×10^9	130×10^9
Human	NA	4×10^9	NA

compartment of the cell, the cytoplasm. In higher organisms (eukaryotes) the DNA is kept in a special organelle, a nucleus, which is bordered by a double-layer membrane. The membrane has many small holes to allow information bearing molecules (DNA, RNA and proteins) to pass between nucleus and cytoplasm.

The information-storing DNA macromolecule is normally a double helix, with two long antiparallel molecules containing the same information in mirror copies in two directions. The two chains are joined together at the nucleotides so that the A pairs with T and C pairs with G. The unidirectional information can be read by the cellular machinery from either strand, in opposite directions. Note that the information is not identical in the antiparallel strands, but complementary, so that the opposite strand message can be obtained by biochemical copying as well as by *in silico* computation. Living cells rely on copying and interconverting this information by the complicated biochemical machinery of DNA and RNA synthesis for cell function and cell division.

In eukaryotes, DNA in the nucleus is normally wound tightly into packages called chromosomes to save space, and unwound only where and when the information is needed. In multicellular eukaryotes (*e.g.* in plants, fungi and animals) there are dozens or even hundreds of chromosomes in the nucleus, and each chromosome can be several hundred million nucleotides long. For example, the human genome (23 pairs of chromosomes) has a total length of about 4 billion nucleotides, coding for some 22,000 genes and corresponding proteins. Each gene can furthermore produce a variety of messenger RNAs (mRNAs) in the RNA processing stage, and thus many different varieties of proteins, called splice variants.

At least half or more of genes have splice variants and alternative polyA sites (termination signals for RNA copying operation), leading to at least 100,000 possible protein variants in a human cell. The complexity of messenger RNAs is increased further by non-translated variants, that is, various mRNA differences in areas which do not code for the protein, for example, alternative lengths of 3' UTRs (untranslated regions after the protein coding segment). Latest research has shown that shorter mRNA UTRs are correlated with increased expression, which may be depend on control by microRNAs (introduced later below). MicroRNAs are known to target mostly 3' UTRs, so shorter UTRs can result in less suppression by microRNAs.

This enormous complexity in the chemical compounds and controlling mechanisms is shown in a simplified form in Figure 1-1 depicting the general structure of a eukaryotic cell.

Duplication of the information in the DNA is performed within a dividing cell by opening the double helix and building another antiparallel strand to each strand of the original helix. This results in two identical copies of the original double helix (barring occasional mistakes in the copying process, causing mutations). The double-stranded helix is more stable than a single molecule, and errors on one strand can be repaired based on the correct information in the other strand.

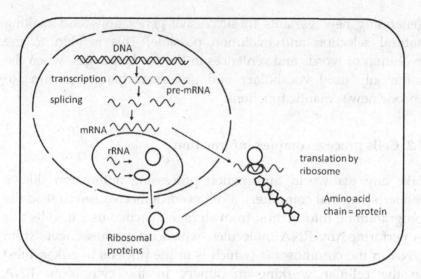

Figure 1-1. Eukaryotic cell structure and DNA transcription and translation according to the main dogma of molecular biology (DNA => RNA => protein). mRNA = messenger RNA carrying the information of coded protein from inside the nucleus to the cytoplasm to the translation site and binds to the ribosomes for translation. rRNA = ribosomal RNA which assembles with proteins to ribosomes, which serve as a scaffold for the translation of RNA to protein.

An additional benefit of the double-stranded molecule is that the information on opposite strands can be used to code for different functions. For example, the meaning of a short string like ATCGGCTTT and its reverse reading AAAGCCGAT on the opposite strand in the opposite direction can encode different meanings for the cell. This efficient packing of information in both directions is utilized by some viruses with very small genomes, when the amount of available DNA is a limiting factor in the life-cycle of the virus.

There can be different types of mistakes, *i.e.* mutations, in the original DNA code, like mistakes in a printed book. The three types of mutations are: transitions (conversions of one or more nucleotides to another alternative of the four nucleotides A, T, C or G), deletions (losses of one or more nucleotides) or insertions (additions of one or more nucleotides). These mutations are essential in generating diversity in the inherited information, thus generating new variants in successive generations and making natural selection and evolution possible. This is akin to the evolution of words and sentences in human languages, when the commonly used vocabulary and grammar changes in time to encode new semantic functions.

1.2 Cells process complex information

Like any real-world information processing automaton (like a modern personal computer), a cell needs a mechanism to read the programming information from storage to active use. In cells this is performed by RNA molecules, which act as messengers from DNA in the chromosome (which is in the nucleus in eukaryotes) to the cellular working machinery in the cytoplasm. RNA molecules are typically quite short, not exceeding a few thousand nucleotides and are less stable than the DNA molecules. Their short lifespan compared to DNA is helpful in producing short-term messages, which are then destroyed, when their data is no longer needed.

The RNA molecules are copies of short messages from DNA, often coding for proteins to be produced. These messenger RNA molecules are then transported from the nucleus through the nuclear membrane holes into the cytoplasm. There the RNA molecules are used to translate the information into specific amino acid sequences to form functional proteins, which run the various machineries for growth, metabolism, cell division and so on.

DNA is first copied to RNA and then triplets of RNA are used to code for the 21 amino acids needed as building blocks for natural proteins. Four nucleotides in triplets can code for 64 messages, so there is ample redundancy in the genetic code and some of the triplets are used for start and stop signals in the production of proteins. Redundancy of the encoding from DNA via RNA to protein means that some mutations do not actually change the encoded amino acid incorporated into the specified protein amino acid chain. These are called silent mutations. Even though such mutations can be silent for the produced protein, they may still have a biological effect. For example, a nucleotide change (transition) may cause DNA-binding proteins to bind differently to the changed DNA, affecting the production of the RNA molecule from the DNA. The change in resulting RNA may also affect the protein production from the RNA due to *e.g.* RNA stability or action of RNA-binding proteins.

This whole system of transcribing of information from DNA to RNA to proteins is called the main dogma of molecular biology, constituting the main direction of the flow of information in the cell from chromosomes to cytoplasm. The functioning of the genetic code is shown in Figure 1-2.

The information flow according to the central dogma is from DNA to RNA to proteins, unraveled in the fifties and sixties. Later research then has found that information flows in various other ways also. Until recently, it has been thought that most of the information in DNA is "junk", as only a very small proportion of sequence codes for proteins. The proportion of such non-coding DNA in various organisms is shown in Figure 1-3.

There is, however, further complexity and feedbacks in the system. For example, RNA molecules can bind to DNA, affecting the DNA function. Special proteins can bind to DNA, also affecting DNA function; an example of this kind of proteins is the class of proteins called transcription factors. These transcription factor proteins move from their production site in the cytoplasm into the nucleus and bind to specific DNA sequences upstream of protein coding regions in the DNA, and control the production of RNAs from the DNA, thus influencing protein production from specific genes. RNA can also bind to proteins, to affect protein function or to form riboproteins, which have many important functions in the cell, like protein translation and chromosome duplication.

Transcription factor proteins were earlier thought to be the main controlling mechanism for gene expression, but recently non-coding RNAs have been shown to be another major layer of

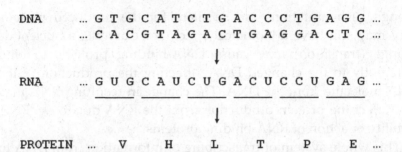

Figure 1-2. Transcription of DNA to RNA and translation to protein according to the genetic code by transcription from DNA to RNA and translation from RNA to amino acid chain of a protein using the triplet code.

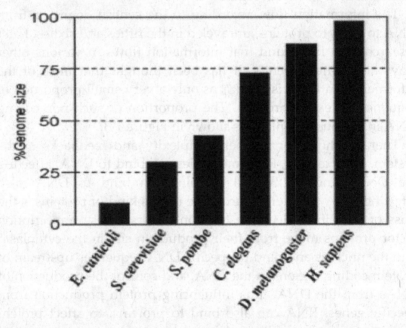

Figure 1-3. Proportions of non-coding DNA in various organisms (Taft *et al.* 2007). Reproduced by permission from John Wiley and Sons.

control, adding to the complexity of the whole system. These non-coding RNAs include a large number of only recently found novel entities (Dinger *et al.* 2008). Examples are shown in Table 1-2.

Table 1-2. Examples of non-coding RNAs, their sizes and functions.

Acronym	Name	Size nt	Function
miRNA	microRNA	21–23	single-stranded, regulate mRNAs
siRNA	small interfering RNA	20–25	double-stranded, regulate mRNAs
snoRNA	small nucleolar RNAs	70–240	guide methylation or pseudouridylation of *e.g.* ribosomal RNAs
ncRNAs	long ncRNAs	>200	regulate gene expression
piRNAs	Piwi-interacting RNA	27–30	germline chromatin modifications and transposon silencing
	Riboswitches	varies	bind small molecules, enzymatic function or modification of mRNA splicing or translation

Especially notable are microRNAs, which are 20-24 nucleotides long tiny RNA molecules that work by binding to messenger RNAs or to genomic DNA to control gene activity. The roles of microRNAs keep expanding, as this is one of the hottest topics in current molecular biology. MicroRNAs have already been demonstrated to be an essential layer of gene expression regulation in addition to small metabolites and transcription factor proteins binding to promoters of genes. It is even considered that small RNAs may be trafficking between cells and act as intercellular messengers in both plants and higher eukaryotic animals (Dinger *et al.* 2008). Non-coding RNAs are also thought to be involved in epigenetic mechanisms in chromatin modification, *e.g.* paramutation (Chandler 2007) and therefore control chromosome function in eukaryotes.

Thus the information processing in cells is very complex and integrates information from DNA, RNA, proteins and the many biochemical compounds of the metabolism. The current view of the network of genetic information flows is in Figure 1-4 below. The role of small non-coding RNAs is becoming increasingly prominent and is even considered to underlie long-term memory (Mercer *et al.* 2008). Interestingly, the importance of non-coding RNA is supported by theoretical calculations of the proportion of regulatory mechanisms needed in a complex metabolic system, based on the lower cost of developing short RNA based control mechanisms compared to the more complicated protein based DNA and RNA control systems (Ahnert *et al.* 2005).

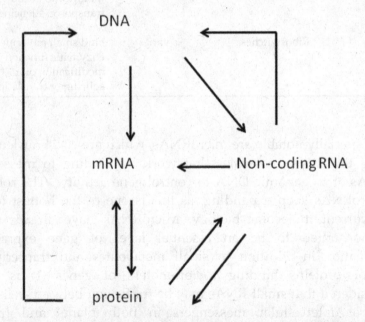

Figure 1-4. Genetic information flows in eukaryotes, including the important role of non-coding RNAs. Proteins bind to mRNAs (*e.g.* ribosomal proteins), to nucleus DNA (*e.g.* transcription factors) and to non-coding RNA (*e.g.* microRNA controlling factors).

1.3 Cellular life is chemically complex and somewhat stochastic

This book is about analyzing the complex web of interactions and information processing that forms the basis of cellular functions. The complexity of life is being studied by an increasing army of researchers, working on DNA, RNA, proteins, metabolism, organic compounds and so on. This has given rise to the large number of "omics" (see Table 1-3). The basic "ome" is the genome as the repository of information from one generation to next, but the so-called epigenome is also now recognized as an essential source of information during the development of a eukaryotic organism and involves DNA modifications, chromatin modifications, histone code, methylation, *etc.*, which do not affect the DNA sequence itself, *i.e.* the sequence of A's, T's, C's and G's (Callinan & Feinberg 2006).

Information storage in cells is more complex than just having a long DNA molecule and a large number of short RNAs floating inside a cell interacting randomly. There are many hardware components necessary to keep the information packaged safely from degradation, and complex mechanisms for its use, as we have seen above, including transcription (DNA to RNA) and translation (RNA to protein). The eukaryotic cell is in fact full of membrane extensions and vesicle systems and macromolecular scaffolds acting as organizing structures for well synchronised and accurately scheduled transport of various materials to their sites of targeted interactions. The complex infrastructure within the cells is evident in the electron microscope pictures of cells showing a large number of vesicles and membrane enclosures.

Genotype variation from one individual to another is a recently emerging complex data domain, when high-throughput methods of recording such variations have become available. Various mutations (transitions, deletions or insertions) in genomic DNA ultimately account for the majority of the metabolic or structural variation in different individuals belonging to the same species. For example, most of the individual differences in humans are due

Table 1-3. Some "omes" of modern biology.

Genome	All genetic information in an organism
Epigenome	Other than DNA-based heritable information
Orfeome	All open reading frames (ORFs) coding for proteins in a genome
Methylome	Pattern of methylated nucleotides in a genome
Transcriptome	All transcribed mRNAs in a cell
Proteome	All proteins coded for in a genome
Interactome	All interactions between macromolecules in a cell
Regulome	All regulatory networks of a cell
Physiome	All physiological functions in a whole organism
Metabolome	All small molecules in a cell in a specific physiological state
Metabonome	Metabolic responses to drugs, environmental changes and diseases.
Glycome	All carbohydrate molecules in a cell
Secretome	All gene products secreted to outside of a cell
Localizome	All locations of various proteins in cell types and subcellular compartments
Unknome	All genes of unknown function

to individual nucleotide mutations, called Single Nucleotide Polymorphisms (SNPs), and sets of such SNPs in limited areas of genome are called haplotypes. The length of the area containing one haplotype indicates the time that has elapsed since the last recombination (crossing over between maternal and paternal chromosomes) in previous generations. The comparative information between haplotypes can be used *e.g.* to track ancestry of mutations and to map disease genes.

In addition to the purely genetic DNA information (*i.e.* the cell's genome or genotype), there exists a large amount of additional epigenetic information based on non-DNA molecules associated with the DNA sequence. The term "epigenetic" means that the underlying DNA sequence can be unmodified, but the additional modifications of chromosomes and their information context is still heritable and can affect gene expression.

Epigenetic mechanisms include various chemical modifications of the DNA, for example methylation and the histone code. The individual DNA nucleotides C and G can be methylated, which affects their message handling (*e.g.* by DNA binding proteins and enzymes processing DNA). The histone code is based on modifications in the histone proteins (mostly acetylation and deacetylation of amino acids), which attach to DNA to package it into supercoiled segments in the eukaryotic chromosomes. Both the DNA methylation and histone modifications are known to be very important in regulating gene expression information. Summary of these various levels of information variants is in Figure 1-5, showing the combination of haplotype and epigenetic variation into so-called hepitypes (= haploepitypes).

Epigenetic complexity is just one facet of the total cellular complexity and variety of regulatory systems in cells. Various other complex data domains in the modern biology era are listed in Table 1-4.

Table 1-4. Examples of complex data domains, which are responsible for the diversity of information processing and molecular interactions in living cells.

DNA replication (in nucleus)	Mutations during DNA replication and RNA transcription
Non-coding RNA	Variety in non-coding RNAs produced
RNA translation machinery in cytoplasm	Splicing of introns/exons to produce a variety of mRNAs
Epigenetics	Histone code, methylation, acetylation
Genetic variation between individuals	Single Nucleotide Polymorphisms as haplotypes
Protein splice variants	Multitude of different proteins arising from variations in mRNAs
Metabolic pathways	A single enzyme can produce a wide variety of end-products from a single precursor, *e.g.* terpenoid synthases
Regulatory pathways, signalling systems	Interactions in the alternative receptor-ligand reactions, microRNAs, transcriprition factors
Cellular fate and tissue differentiation pathways	Variety of developmental pathways, *e.g.* in development of blood cell types from stem cells.

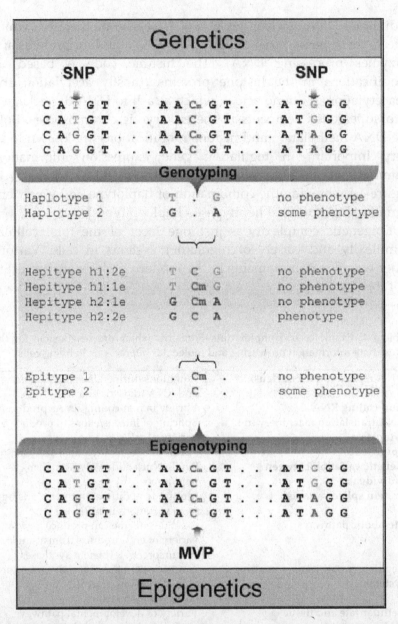

Figure 1-5. Relationships of genetic and epigenetic variations in eukaryotic cells. (Murrell 2005). Reproduced with permission from Oxford University Press.

Our information about cellular life has also stochastic aspects. As in any large and complex physical system, there are errors and random variations in basic life processes. Correct understanding of biological systems is further obscured by inaccurate data obtained from our laboratory experiments.

However, nature has produced some amazingly accurate biological mechanisms for maintenance of biological information. The most accurate mechanism by far is the DNA replication to copy the genetic information between dividing cells. Its accuracy has been developed during hundreds of millions of years to a degree, which is envy for modern nanotechnologists working on controlled molecular assembly and nanomotors. A multitude of enzymes and large protein complexes are involved in unwinding the double helix of DNA and synthesizing the new strand of DNA in the replication fork (see Figure 1-6, top). This process reaches average synthesis accuracy of 10^{-7} to 10^{-8} (Kunkel 2004) both in living cells *in vivo* and in test tubes using isolated DNA replicating enzymes *in vitro*. This is still further improved by several repair mechanisms that make corrections after copying, leading to an overall accuracy of up to 10^{-9} in eukaryotic multicellular organisms.

Local variations along the genome in the fidelity of translation occur widely, especially in the so-called hotspots, where mutations are very abundant, and cold spots, which are more carefully kept intact. These have important roles for the evolution of organisms, both to develop new variation and new genes and to preserve crucial function unaltered, when necessary. Thus evolution has found ways of utilizing stochasticity to its advantage, when new mutations and new information needs to be created for the needs of the organism to adapt to new environments by modifying its genome in successive generations.

The DNA is not copied simply continuously from one end to the other on both strands. One strand of the double helix is 5' to 3' based on the orientation of the sugar backbone, and the other strand is 3' to 5', in an antiparallel orientation. New DNA copying

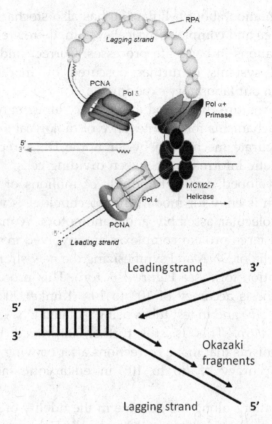

Figure 1-6. Schematic diagram of the protein complexes which control DNA replication (top, McCulloch & Kunkel 2008) and the replication of leading and lagging strands of DNA (bottom). Reproduced with permission from Nature Publishing Group.

operation by DNA polymerase only works in the 3′to 5′ direction, so that synthesis is continuous in one direction (3′to 5′direction, leading strand), but occurs in bursts of about 200 base pairs (called Okazaki fragments) in the other, lagging strand, as shown in Figure 1-6 (bottom).

Due to the multiple starting points in the Okazaki fragments, there are more possibilities of things going wrong in the lagging strand, due to more enzymatic operations, and joinings of DNA pieces together. Indeed, in bacteria the lagging strand mutations are 3- to 10-fold more abundant than mutations in the leading strand (Fijalkowska *et al.* 1998) and in yeast, 2- to 8-fold (Pavlov *et al.* 2002). Even in the leading strand, the DNA in mammalian chromosomes is not completely continuous either. DNA synthesis in a medium-sized human chromosome for example is initiated in up to 40,000 locations in the leading strand (at so-called replication origins), but in the lagging strand initiation occurs at up to 20 million locations (Hubscher *et al.* 2002).

In bioinformatics analyses, it is useful to bear in mind these basic biological sources of nucleotide data variation. Also, the current DNA sequencing methods often have much larger experimental error rates than the natural accuracy of DNA replication of 10^{-7} to 10^{-8}. Modern DNA sequencing machines (including the new Roche 454 GS20 sequencer introduced in 2006) can have sequencing error rate of 10^{-3} to 10^{-4}, depending on equipment and protocols. When sequencing is done repeatedly and results pooled in multiply aligned sequences, a much more reliable consensus sequence can obviously be built, provided reliable copies of the same molecule are at hand to start with.

Already in 2003 the most accurate human genome assembled genomic sequences from Sanger Centre in UK had a reported error rate of only 10^{-5} per nucleotide (Anonymous 2003), using the traditional Sanger method sequencing machines. Note that current methods rely on sequencing of multiple copies of the same molecule. Novel technologies relying on single-molecule sequencing are likely to become available in a few years, possibly increasing the accuracy of primary data. Until that time though, detailed comparisons of single nucleotide differences between DNA sequences need to take into account both the possible errors in biological replication and in *in vitro* amplification of the original DNA molecules that have been used as well as the experimental errors occurring in the sequencing machine and sequence assembly stage.

Thus it is good to remember both the stochastic nature of the "natural" errors present in imperfect operation of cellular machines and the experimental errors in any biodata acquisition.

One should not try to attempt to predict things more accurately than what happens in nature or is warranted by the accuracy of the raw biodata inputted into the prediction system or algorithm.

1.4 Challenges in analyzing complex biodata

As we have seen in this cursory introduction to the complexities of cellular life, the biological systems are akin to very complex physical systems, like weather systems or galaxy formation, with the difference that the number of types of interactions is larger and that there are many simultaneous interacting chains of deterministic causalities which makes the cellular life so efficient.

In nanoscale the biological systems can have strong randomizing processes (including Brownian motion and random destruction of biomolecules by oxidants) affecting how individual molecules move and function within cytoplasm. However, in a slightly larger scale relevant to living cells the overall functions are surprisingly deterministic. The challenges for biodata analysis lie both in selecting the relevant data from the correct physical scale, as well as measuring the correct output and dependent variables for the problem in hand.

The other main issue facing bioinformaticians is the reliability of experimental data and even metadata (*e.g.* names and functions of genes, proteins *etc.*). Even the best algorithms cannot retrieve the correct conclusions from summarized and integrated multi-domain data, if the underlying raw data has unquantified uncertainties. Therefore proper sensitivity analysis for possible natural variation in the raw data is always warranted.

References

Ahnert, S.E., Fink, T.M.A. & Zinovyeva, A. 2008. How much non-coding DNA do eukaryotes require? *Journal of Theoretical Biology 252(4): 587–592.*

Anonymous. 2003. Wellcome Trust, Sanger Center Press Releases: 14th April 2003. The Finished Human Genome — Wellcome To The Genomic Age. http://www.sanger.ac.uk/Info/Press/2003/030414.shtml *(accessed 8 July 2007)*

Callinan, P.A. & Feinberg, A.P. 2006. The emerging science of epigenomics. *Human Molecular Genetics 15(1): R95–101.*

Chandler, V.L. 2007. Paramutation: from maize to mice. *Cell 128: 641–645.*

Dinger, M.E., Mercer, T.R. & Mattick, J.S. 2008. RNAs as extracellular signaling molecules. *J Mol Endocrinol. 40(4): 151–159.*

Fijalkowska, I.J., Jonczyk, P., Tkaczyk, M.M., Bialoskorska, M. & Schaaper, R.M. 1998. Unequal fidelity of leading strand and lagging strand DNA replication on the *Escherichia coli* chromosome. *PNAS USA 95: 10020–10025.*

Forster, A.C. & Church, G.M. 2006. Towards synthesis of a minimal cell. *Molecular Systems Biology 2: 45.*

Gregory, T.R. 2006. Animal Genome Size Database. *http://www.genomesize.com.*

Hübscher, U., Maga, G. & Silvio Spadari, S. 2002. Eukaryotic Dna Polymerases. *Annual Review of Biochemistry 71: 133–163.*

Kunkel, T.A. 2004. DNA Replication Fidelity. *J. Biol. Chemistry 279: 16895–16898.*

Mattick, J. S. 2007. A new paradigm for developmental biology. *Journal of Experimental Biology 21: 1526–1547.*

McCulloch, S.D., Kunkel, T.A. 2008. The fidelity of DNA synthesis by eukaryotic replicative and translesion synthesis polymerases. *Cell Research 18: 148–161.*

Mercer, T.R., Dinger, M.E., Mariani, J., Kosik, K.S., Mehler, M.F & Mattick, J.S. 2008. Noncoding RNAs in Long-Term Memory Formation. *The Neuroscientist 14(5): 434–445.*

Murrell, A., Rakyan, V.K. & Beck, S. 2005. From genome to epigenome. *Human Molecular Genetics 14(1): R3–R10.*

Pavlov, Y., Newlon, C. & Kunkel, T. 2002. Yeast Origins Establish a Strand Bias for Replicational Mutagenesis . *Molecular Cell, 10(1): 207–213.*

Taft, R.J., Pheasant, M. & Mattick, J.S. 2007. The relationship between non-protein-coding DNA and eukaryotic complexity. *BioEssays 29(3): 288–299.*

Zonneveld, B.J.M., Leitch, I.J. & Bennett, M.D. 2005. First nuclear DNA amounts in more than 300 angiosperms. *Annals of Botany 96: 229–244.*

Zubáčováa, Z., Cimbůrekb, Z. & Tachezya, J. 2008. Comparative analysis of trichomonad genome sizes and karyotypes. *Molecular and Biochemical Parasitology 161(1): 49–54.*

Chapter 2

Biodata Explosion

The scope of biodata domains spans all of the biological complexity covered in the introduction. Each eukaryotic cell contains a large amount of information in its DNA, and hundreds of thousands kinds of RNAs produced from the DNA. The RNAs in turn produce tens of thousands basic types of proteins, but many variations of each protein exist, so that the number of protein variants also number in the hundreds of thousands. In addition to the basic databases cataloguing known DNA, RNA and protein sequence data, there exists a large number of secondary annotation databases, which contain DNA, RNA and protein motifs, gene function annotations and phylogenetic relationships of the genes to similar genes and gene families in other organisms.

A good annual review of molecular biology databases appears in the beginning of each year in the journal Nucleic Acids Research, which maintains a list of recognized databases. In 2009 the list contained 1,170 databases; the list is available at http://www.oxfordjournals.org/nar/database/a/.

These above-mentioned datasets are still only the scaffold of data structures, to which the large and fast increasing collections of experimental quantitative data are linked. The experimental datasets include microarray data on expression level of mRNAs in cells, proteomic array data on protein expression levels, and general metabolomics data on other biochemicals in cells and organisms, as well as data from drug development experiments and medical trials.

On a higher level, there are also huge datasets accumulating which contain information about genetic variations between individuals of the same species, *e.g.* single nucleotide polymorphisms (SNPs) of humans, plants and animals. Further in the progression toward metadata (data about data) are databases concerning interactions between proteins, RNAs and metabolites, which is the domain of systems biology. At the highest level are then databases about biological ontologies and charts of metabolic pathways, as well as text-mined associations between biological processes and research publications.

This chapter reviews the gamut of currently available databases and their complex contents, starting with primary data on DNAs and proteins, followed by annotation, experimental and metadata databases.

2.1 Primary sequence and structure data

2.1.1 *DNA sequence databases*

2.1.1.1 *Gene sequence data*

The main repositories of DNA and RNA primary sequences are historically the NCBI GenBank, EMBL and DDBJ databases in USA, Europe and Japan, respectively. Note that RNA sequences (having U nucleotide instead of T, as in DNA) are normally represented by the corresponding DNA sequence. These databases are freely available on the internet and are regularly cross-referenced and kept mutually up-to-date to contain essentially the same data, but the data is organized slightly differently. These databases also include the translations of protein coding segments of RNAs to protein sequences, though separate comprehensive protein databases also exist and are described later below.

In October 2009 NCBI GenBank contained over 110 x 10^6 records, consisting of both short partial and full sequences of genes and partial sequences of genomes, totalling a staggering 257 billion nucleotides of data, which amounts to about 50 gigabytes of data. This corresponds to the data encoded in about half a dozen DVD movies. A short tabulation of NCBI GenBank data is in Table 2-1 below.

The historical background of these databases is that initially the individually studied genes were partially or fully sequenced and deposited one by one to the databases. Later on, sequencing technology advanced and large-scale genome projects submitted partial genome fragment sequences as well, as primary data for early data release.

Gradually the genome sequence projects assembled full genomes, which then became separate sections in these databases, but the original genome fragments are still in the primary data repositories, including even the historical versions of earlier submitted sequences. Therefore the primary unfiltered and unprocessed databases contain a lot of duplication and redundancy which may complicate data analysis. Specially curated non-redundant sections of both the DNA and protein sequences are also offered by the main databases to speed up similarity searches and reduce the storage requirements of databases, subsets of which may be stored locally at the scientists' own computers for local processing.

Table 2-1. NCBI GenBank database contents as of October 2009.

Dataset	Database entries	Total base pairs
Whole genome sequences	48 x 10^6	149 x 10^9
Subset: Human genome	27 x 10^3	3 x 10^9
Others (genes etc.)	111 x 10^6	108 x 10^9
Total:	159 x 10^6	257 x 10^9

In addition to the huge global DNA databases listed above, there are numerous more specific DNA databases dedicated to specific biology domains and/or species. RNA databases are also becoming available, due to the recognition of the special importance of RNA molecules as opposed to DNAs. For example, RFAM database (Griffiths-Jones *et al.* 2008) aims to catalogue RNA families of non-coding RNAs, cis-regulatory elements and self-splicing RNAs with multiple sequence alignments and consensus secondary structures.

2.1.1.2 *Species-specific genome data*

With the advent of faster DNA sequencing technologies and efficient software for genome assemblies, the count of sequenced prokaryotic and eukaryotic genomes, both assembled finished genomes and unassembled ongoing genome projects is still growing exponentially. Recently a new DNA bar-coding based method of using high-throughput sequencers for simultaneous genome mapping and assembly was proposed which promises to make genome sequencing and assembly a truly routine task (Jiang *et al.* 2009). The current GOLD database (Genomes OnLine Database at http://www.genomesonline.org/, Liolios *et al.* 2008) contents are given in Table 2-2 below.

Table 2-2. Genomes OnLine Database contents in October 2009.

Group	Sequenced	In progress	Total
Eukaryotes	119	1,207	1,326
Archaeal	70	111	181
Prokaryotes	940	3,444	4,384
Total	1,129	4,762	5,891

While the big sequencing centres offer large and well annotated genome databases, they are not all made the same way, nor is the data formatted uniformly. This is even more evident in smaller specific genome databases. There is a standardization body (Genomics Standardization Consortium), which tries to improve integratability of various genome databases and Guidelines for Minimum Information about a Genomic Sequence (MIGS) have been published (Field *et al.* 2008). However, the onus is on the bioinformatician to scrutinize the data formatting standards carefully before merging data from different databases.

Recently there has been also a suggestion to do the same for published experimental data, so that the data generation methods, downloadable data sets and use of results would be in a standardized machine-readable form (Garrity *et al.* 2008).

2.1.1.3 *Metagenome data*

There are several separate metagenomic websites, but most such datasets are available in NCBI. As of August 2008 the NCBI database comprised a total of 211 metagenomic datasets, the majority from environmental metagenome sequencing. Some examples illustrating the diversity of available and quickly accumulating data are in the Table 2-3 below.

Table 2-3. Examples of metagenome datasets in NCBI in 2008.

human gut microbe	mine drainage biofilm microbes
bovine gut microbes	wastewater treatment plant microbes
chicken gut microbes	beach sand microbes
termite gut microbes	fresh water viruses and microbes
fish gut viruses	coastal marine bacterioplankton
fish slime microbes	marine bacteriophages from Arctic Ocean
mosquito viruses	hydrothermal vent epibionts
coral microbiome	hot spring viruses

There are some 130 metagenome projects in progress (Anonymous 2008a) and more are being planned. An important recent project is the International Human Microbiome Consortium launched in October 2008 which joins the US NIH data with the European Union METAHIT data (Anonymous 2008b), and plans to sequence a large number of individual human gut microbiomes.

Some datasets have been carefully assembled to contigs (continuous long DNA sequence segments joined from overlapping subsequences) to include many full protein coding bacterial and viral sequences and even large genome segments from individual microbes living in the sampled environment. These datasets are important biodiversity samples in their own right, and can include DNA and protein variants from large numbers of individual microbes or viruses of a single species. Therefore these data can give an interesting glimpse of evolution at work by showing the range of surviving mutations in large populations of microbes in their natural setting, which could not be obtained in even the largest scientific laboratories. In fact, many of the sampled microbes in these metagenomic data are even unculturable in the laboratory and may be only known from their DNA signatures from the metagenomic samples.

Data presentation standards for metagenome data are also being developed by the Genome Standardization Cooperative, but again the onus is on the bioinformatician to consider carefully the data formatting standards before merging data from different metagenome projects.

Within the UniProt protein database (which tries to encompass all known protein sequences in a non-redundant manner) there is also a special section containing translated protein sequences from metagenomic and environmental DNA sequences (like the famous Sargasso Sea water sample sequences and soil and animal gut sample genome sequences *etc.*). One big dataset is the Global Ocean Sampling (GOS) dataset, which contains some 25 million sequences with over 6 million predicted protein sequences from an unknown number of marine microorganisms.

2.1.1.4 *Multiply aligned genome datasets*

At the next data integration level in genomics, multiple alignments of several bacterial or eukaryotic species have become an essential tool for comparative genomics. Bacterial multiple genome alignments are more easily accomplished, due to the smaller genome sizes (1–3 million bases each), and many sets of aligned bacterial genomes are now available from numerous research groups on the internet. For eukaryotic genomes, the necessary resources are much larger, and few sites offer comprehensive multiple genome alignments. The main such site is the European Bioinformatics Institute ENSEMBL website (Anonymous 2008c), which has produced several global multiple alignments of vertebrate genomes, in addition to many pairwise alignments between *e.g.* human and chimpanzee, rat and mouse *etc.* The main datasets as of August 2008 include one vertebrate alignment of 12 species from *Platypus* to rat to cow to human (so-called Pecan analysis), one mammalian alignment of 23 species (EPO analysis) and a four-species alignment of human and three monkey species.

2.1.2 *Protein sequence databases*

Protein databases contain both amino acid sequences from real isolated proteins and an increasing amount of *in silico* derived translated amino acid sequences from mRNA and genomic DNA sequences. These databases are therefore derived data, with concomitant caveats for correctness and quality. Real proteins may in fact differ from the proteins derived from DNA or RNA sequences, because post-translational modifications are common in proteins after their synthesis in cytoplasm. Such modifications include deletions of peptide domains or deletions and modifications of single amino acids.

All the major DNA databanks (NCBI's GenBank, EBI's EMBL and Japan's DDBJ) contain in addition to their DNA data, corresponding sets of proteins. Databases targeting specifically proteins include the European SwissProt (now part of EBI UniProt), PDB (Protein DataBase) in USA and PDBj in Japan. The SwissProt has expanded to UniProt, which now encompasses all proteins resulting from the collaboration between the European Bioinformatics Institute (EBI), the Swiss Institute of Bioinformatics (SIB) and the Protein Information Resource (PIR). In March 2009 UniProt release 40 of UniProtKB/TREMBL contained sequences for some 7.7 million entries with over 2.5 billion amino acids, mostly proteins translated from DNA. A good introduction to UniProt can be found in Mulder *et al.* (2008), which also describes the protein motif database InterPro and a cross-linking database Integr8, discussed below.

2.1.3 *Molecular structure databases*

2.1.3.1 *Protein structures*

Protein structural databases have recently joined together to provide web-based database serviced named Worldwide Protein DataBank (Anonymous 2008d). In addition to the US, European and Japanese protein databanks, it also includes Biological Magnetic Resonance Data Bank (BMRB) which comprises nuclear magnetic resonance (NMR) spectra of proteins, peptides and nucleic acids. In 2008 the database contained over 49,000 entries for molecular structures obtained from analysis of crystallized proteins or from *in silico* analysis. The database entries list the structures with atomic coordinate data in the legacy PDB flat file data format, though a modified relational database format is available for querying or downloading from EBI. Recently, the database has been restandardized (Henrick *et al.* 2008) and now represents a high-quality source of various annotated data as well. The annotations include standard chemical compound nomenclature, such as SMILES and IUPAC International Chemical Identifiers for the subunits of proteins and ligands bound with proteins for cocrystallized molecules.

2.1.3.2 *Chemical structures*

Non-protein molecular structures also have their own massive databases, most important being PubChem (Xie & Chen 2008). PubChem includes three main interlinked databases: PubChem Substance, PubChem Compound, and PubChem BioAssay. PubChem Substance contains over 47 million records (December 2008) of chemical samples from various depositors, including chemical vendors and academic institutions. The chemical samples are then linked to PubChem Compound database, which contains validated chemical structures of pure chemical compounds. The number of chemical structures currently contains over 19 million unique structures, which have been pre-clustered and cross-indexed to hierarchical similarity groupings for ease of access to related compounds. One of the most useful tools in the Compound database is clustering tree representation of compound similarities. Finally, PubChem BioAssay includes information about experimental bioactivity results for any of the substances in PubChem Substance database. Currently over 1,000 bioassay experiments are in the BioAssay database.

For small non-protein molecules there is also crystallographic data on some 436,000 compounds at the Cambridge Structural database (CSD, Anonymous 2008e). There are also many commercial databases, some quite expensive, offering similar services, CAS (Chemical Abstract Services) being the prominent example.

RNA structural databases are also starting to appear for secondary folding conformations of small RNAS, for example, RNA STRAND version 2 contains about 5,000 folding structures of various RNAs, including mRNAs, ribosomal RNAs, ribozymes and so on (Andronescu *et al.* 2008, http://www.rnasoft.ca/strand/).

2.1.3.3 *Epigenomic data (methylation, histone codes)*

Methylation of DNA is an ancient mechanism to protect DNA from DNA cutting enzymes called restriction enzymes, which are used for protecting the cell from unwanted foreign DNA, *e.g.* invading viruses, for regulating gene expression and likely also for controlling mobile genetic elements called transposons. Thus methylation status of the genome is very important for the integrity of genetic information, and has many other essential functions. The evolutionary history and functions of DNA methylation are well reviewed in Zilberman (2008).

Until recently, little information on genome methylation was available. The main efforts in large-scale epigenomics have so far been in high-throughput sequencing of methylated individual nucleotides in the human genomic DNA, spearheaded by the Human Epigenome Project (HEP) at Sanger Centre (Anonymous 2008f). Their website already gives access to some 2 million methylation values from human chromosomes 6, 20 and 22, spanning selected areas of three human chromosomes. Technology for the sequencing of methylated nucleotides in the genomic DNA is evolving fast, and ultra-high throughput sequencing methods are becoming prominent in this area (Zilberman & Henikoff 2007), so that larger databases are likely to become available soon. In addition, many small specialised methylation databases contain information about disease-specific genes and their methylation variants in different individuals. Already a single-base pair resolution methylation map is available for the *Arabidopsis* model plant (Cokus *et al.* 2008).

Recent bioinformatic evaluation of accumulated DNA methylation data has shown that in mammals the genomic DNA areas which are rich in CG dinucleotides (so-called CpG rich areas) have quite consistent average methylation rate of the cytosine (C). These areas are thus not very informative for inter-individual differences. However, it has been shown that the CpG-poor areas, in which small-scale high-resolution methylation patterns have been found, may be important in determining inter-individual differences in physiological function and susceptibility to disease (Bock *et al.* 2008). For example, it has already been found that methylation patterns underlie schizophrenia symptoms in human patients (Schumacher *et al.* 2008). As we will often see later on in this book, a single biological factor alone, like the methylation pattern, however, often does not explain all of the inter-individual differences, so that data from different aspects of metabolism and signalling has to be combined. As an illustration of this, integrated analysis combining standard microarray expression analyses with methylation profiles of selected genes has been shown to more efficiently discover and evaluate leukaemia-related genes and their variation in different individuals having two different types of leukaemia disease (Figueroa *et al.* 2008).

As for the histone code, many mechanisms of histone modifications are already known. For example, histone acetylations H3K9ac (meaning acetylation at position 9 amino acid of histone protein H3K) and H4K16ac and histone methylation H3K4me signal active DNA transcription, while histone methylations H3K9me and H3K27me signal repression of DNA transcription. High-resolution data is also becoming available, for example, *Arabidopsis* chromatin histone methylation has been analyzed at high resolution up to nucleosome level (one nucleosome protein complex is a set of histones wrapping around the compacted DNA and spans about 147 nucleotides). Interestingly, histone modification appears strongly correlated with DNA cytosine methylation (Bernatavichute *et al.* 2008).

2.2 Secondary annotation data

In addition to the explosion of primary DNA and protein sequence data, a similar explosion has occurred in secondary annotation data, which encompasses all additional experimental or deduced information from combined data sources linked to primary identified DNAs and proteins. At a more detailed level, annotations are information about subsequences in the DNAs and proteins in the form of various motifs, or fingerprints of the salient features which have been shown to be functionally relevant in one way or another in metabolism or signalling. Other types of annotations pertain to the whole gene or protein, *e.g.* experimental indications about the biological role of the DNA or protein. At a higher level still, the whole genome itself can be annotated for subsequences of various lengths of the genome, which can even be overlapping each other.

All these kinds of annotations are to some degree included in the main international DNA and protein sequences databases, either as embedded data in the sequence accessions, or, nowadays more commonly as links to other databases of annotated features. At the genome scale, these are known as annotation tracks along the genome, and can be easily added and visualized in *e.g.* the ENSEMBL genome visualisation website.

2.2.1 *Motif annotations*

Motifs are generally short sequence patterns denoted by specific conserved nucleotides (A, T, C or G) at each position of aligned DNAs or proteins, including specification of gap and insertion location lengths. The IUPAC letter codes for DNA/RNA nucleotides and protein amino acids are standardized and universally utilized as a standard text string alphabet for such motifs. Occasionally some smaller specialized databases use slightly different conventions, which is good to keep in mind when merging such data together with the larger standard motif databases.

2.2.1.1 *DNA motif databases*

An important database for short motifs of DNA binding sites utilized by transcription factor proteins is TRANSFAC (Matys *et al.* 2006). The last freely accessible version 7.0 from 2005 contains about 8,000 binding motifs as simple consensus sequences for 1,500 genes and over 6,000 transcription factors. Other notable

DNA motif databases include JASPAR for eukaryotic transcription factor binding profiles (Sandelin *et al.* 2004) and SCPD for yeast promoter motif database (Zhu & Zhang 1999). In addition, there is a large number of specialized small databases for motifs restricted to certain organisms or functions of the motifs, like transcription start signals, promoter motifs *etc.* Many of these can be found in the annual January database issue of Nucleic Acids Research journal.

2.2.1.2 *Protein motif databases*

The best known traditional motif annotation system is the ProSite motif developed by Amos Bairoch (Sigrist *et al.* 2002, 2005). It is available at http://expasy.org/prosite/. It is a good example demonstrating the methods of generating and using **patterns** and **profiles**. A **pattern** is a fixed sequence string containing listings of allowed amino acids at each position. A **profile** is a data pattern giving the frequency or probability of each kind of nucleotide at each position of the motif.

ProSite patterns can be expressed as computer-parseable search strings in a linear format. For example, the signature of the C2H2-type zinc finger domain is written in ProSite syntax as follows:

```
C-x(2,4)-C-x(3)-[LIVMFYWC]-x(8)-H-x(3,5)-H
```

where x(2,4) means from 2 to 4 amino acids of any kind, x(3) means 3 amino acids of any kind, [LIVMFYWC] means one amino acid selected from the list in square brackets, and so on.

This pattern representation does not capture all the available information from a multiple alignment of many sequences, because the pattern regards all variants equally likely, regardless of the frequencies of each amino acid at each position in the actually observed set of sequences.

For this purpose, the ProSite **profiles** include more information by giving the position frequency matrix recording the frequency of each amino acid at each position. In essence, this is a statistical description of the multiple alignment of sequences that are considered to belong to the group having the motif. This matrix can then be used to calculate a position weight matrix containing log-odds weights for computing a matching score for a specific amino acid sequence.

The general ProSite profile representation is similar to that used in Hidden Markov Model (HMM) profiles, *e.g.* by HMMER software, and such profiles can be interconverted, for example from HMM to ProSite by program htop using the tool pftools (http://www.isrec.isb-sib.ch/ftp-server/pftools/pft2.3/).

The latest development is the ProRule database, which is a machine-readable manually curated collection of rules for identifying patterns based on ProSite motifs (Sigrist *et al.* 2005). It offers further sensitivity and specificity in identifying specific conserved motifs. The ProSite team has also made available a search tool ScanProSite, both as a downloadable executable as well as a web service. The main advantage of motif/pattern based searches is the speed of the algorithms, because for example a complete search of ~4,000 proteins in *Escherichia coli* bacterial genome can be searched by one standard PC CPU in 10 minutes. In the end of 2007 ProSite database contained about 1300 patterns, 750 profiles and 750 ProRules.

Other notable protein motif databases include SMART (Bateman *et al.* 2008), BLOCKS (Henikoff *et al.* 2000) and PRINTS (Attwood *et al.* 2003). However, most of the important motif databases have been usefully integrated in a single very useful motif resource called INTERPRO (Hunter *et al.* 2008), which currently integrates PROSITE, PRINTS, Pfam, ProDom, SMART, TIGRFAMs and others. Currently about 24,000 motif signatures are contained within INTERPRO (version 18.0, September 2008) out of the total of 57,000 signatures in the participating databases. One big advantage is the InterProScan motif search tool that is available as web service or as standalone downloadable application to search your own sequences for known InterPro motifs.

2.2.2 *Gene function annotations*

The main gene function annotation system for genes coding for proteins is the Gene Ontology database, available at GeneOntology.org. It is a controlled vocabulary and annotation database for species-independent gene functions. Species-specific subsets of data are also available for human, mouse, rat, zebrafish, chicken, cow and *Arabidopsis* proteins. Each annotation for each protein is also given an evidence type code as follows:

IMP = inferred from mutant phenotype
IGI = inferred from genetic interaction
IPI = inferred from physical interaction
ISS = inferred from sequence similarity
IDA = inferred from direct assay
IEP = inferred from expression pattern
IEA = inferred from electronic annotation
IGC = inferred from genomic context
TAS = traceable author statement
NAS = non-traceable author statement
ND = no biological data available
IC = inferred by curator
RCA = reviewed computational analysis

This system allows people (and software) to filter the data according to the evidence they deem suitable for the analysis at hand. For example, IDA (inferred from direct assay) is considered by some to be the most reliable evidence type, while IEA (inferred from electronic annotation) may occasionally be incorrect. One version of the database is stored in EBI GOA database at http://www.ebi.ac.uk/GOA/. The GOA database is described in detail in Barrell *et al.* (2008).

The complete GOA annotation database for UniProt in April 2008 included over 4.5 million proteins with over 34 million gene ontology label associations to hundreds of thousands of genes (one protein can have several annotations at hierarchical level of accuracy and even several distinct functions). This large dataset links gene function annotations to most UniProt items that have sufficient similarity to proteins of known function. Obviously, most of the annotations (about 98%) are based on electronic annotations or computational analyses based on a relatively small amount of experimental data from model organisms.

2.2.3 *Genomic annotations*

All major suppliers of public domain genome data also have annotations to the sequences at different levels of resolution. Each type of annotation is termed an annotation track, and the various annotation tracks may be visualized below the zoomable genome sequence in various ways and specific sets of tracks may be chosen according to the user.

ENSEMBL genome database has a sophisticated annotation track system (www.ensembl.org), currently containing some 150 annotation tracks. The NCBI genomic display MapViewer (http://www.ncbi.nlm.nih.gov/mapview/) shows similarly zoomable genome sections with selectable genome annotations. The amounts of data are massive, and can only be browsed in very limited subsets by the available visual interfaces. The other main genome viewer to access genomic annotation data is the UCSC Genome Browser at http://genome.ucsc.edu/index.html.

In general, sequence based annotation systems are becoming very complex automated computing workflows and a good discussion of the present state of art can be found in Juncker *et al.* (2009).

2.2.4 *Inter-species phylogeny and gene family annotations*

Phylogenetic analysis is an extensive research area and many annotation databases based on interspecies similarities and gene families have been developed for specific gene families or taxonomic groups. Taxonomy of living organisms for the at least 2 million species on earth is finally also moving to the internet, as many biodiversity recording and organizing databases have appeared recently and taxonomic e-science is emerging (Godfray *et al.* 2007, Mayo *et al.* 2008). However, for the majority of DNA and protein databases, the NCBI TaxBrowser classification is the *de facto* standard for naming of organisms, although it is not a fully standardized taxonomic database approved by professional taxonomists or the International Code of Zoological Nomenclature and other codified traditional taxonomic nomenclatures. Currently (end of 2009) the TaxBrowser includes about 212,000 species labels attached to the submitted DNA and protein sequence accessions in the NCBI database.

Most DNA and protein motif databases discussed above include links to phylogenetic relations and taxonomy of the species, most commonly via the TaxBrowser section of the NCBI. NCBI also has a comprehensive Conserved Domain Database (CDD), which covers all the non-redundant protein sequences deposited in NCBI (Marchler-Bauer *et al.* 2008). Conserved protein domains are clustered using reverse position-specific BLAST, in which the query sequence is compared to a position-specific score matrix (PSSM) prepared from the underlying conserved domain alignment. In early April 2010 there were some 40,000 protein domains in the database, and a new tool (called CDTree) for browsing the hierarchy tree of related domains is available as web service or as standalone tool for protein family analysis.

Truly global integrated views of DNA and protein phylogenies for all genes in all organisms do not exist yet, but many more limited domain-specific databases with comparative data from different species are already available. One of the more extensive ones is the COG (Clustered Orthologous Groups of proteins) database at NCBI, which currently includes clustered protein families from 66 unicellular organisms (bacteria and fungi).

An eukaryotic clustering from 7 species (the KOG database) is also provided separately, and the included species in both databases keep increasing (Tatusov *et al.* 2003). Recently, NCBI launched also a clustered protein database ProtClustDB, which clusters 1.7 million proteins from 3,800 genomes of prokaryotes, bacteriophages (viruses of bacteria), mitochondria and chloroplasts (Klimke *et al.* 2009). It is based on the NCBI RefSeq sequence set, which is a non-redundant collection of all genes and proteins submitted to NCBI.

An example of a domain-specific motif database is SWISS-REGULON, which contains genome-wide annotations of regulatory sites in intergenic genomic DNA of yeast and 17 different prokaryotic genomes. The annotations are based on combined data from known regulatory motifs from the literature, *in silico* prediction from similar sites in other species and data from various microarray experiments (Pachkov *et al.* 2007). Another example of a domain-specific DNA motif database is MiGenes, which is a gene ontology annotation based interspecies database of mitochondrial proteins and their conserved protein motifs (Basu *et al.* 2006).

2.3 Experimental and personalized data

The next level of secondary annotation relates genomic features to inter-species evolutionary history and to differences between individuals of the same species, *e.g.* personalized medicine. Here the new high-throughput sequencing machines and other nanotechnology based high-throughput experimental devices will transform the biological research in a fundamental way. In the following the current main types of experimental datasets are briefly reviewed.

2.3.1 *DNA expression profiles*

DNA microarrays is a powerful tool to measure thousands and even hundreds of thousands of different mRNAs corresponding to expressed proteins from isolated RNA samples from tissues and cells. Microarrays consist of large arrays of known DNAs attached at known locations at a very high density on a solid surface. The DNA sample to be analyzed is poured on the array, hybridization between array DNAs and sample DNAs is effected and the hybridization signal is captured to get a measure of the abundance of the specific mRNAs in the samples. A voluminous literature exists on standardization of acquisition and analysis of microarray data. Some algorithms in this area will be reviewed later in this book with illustrative examples of novel approaches.

Extensive amounts of microarray data of varying quality and complexity appears in numerous project specific websites around the internet. Increasingly the submission of data to databases is in standardized MIAME and other microarray related data formats prepared by the Microarray and Gene Expression Data (MGED) Society (http://www.mged.org/) and is becoming routine in larger microarray projects. NCBI is collecting a large repository of such data in its Gene Expression Omnibus (GEO) database (Barrett *et al.* 2009), which currently contains over 270,000 samples analysed in ArrayExpress (Parkinson *et al.* 2008) and also includes a quarter of a million hybridizations. GEO contains a variety of omics data in addition to standard microarray data, including ArrayCGH, SNP Arrays, Serial Analysis of Gene Expression (SAGE), Massively Parallel Signature Sequencing (MPSS), protein arrays, and mass spectrometry data.

In addition to the global microarray data repositories, there are many specialized microarray databases. One worth mentioning is the non-coding RNA expression database NRED (Dinger *et al.* 2009) which reflects the growing importance of non-coding RNA research.

Microarray technology is likely to be overtaken in many applications by the new ultra-high throughput DNA sequencing methods, because soon from a technical point of view one will be able to sequence literally any DNA one wishes in a biological experiment (Kahvejian *et al.* 2008). This will mean that the quantitative (analog) data about expression levels of genes and amounts of mRNAs corresponding to proteins in cell populations is expected to become digital information of exact numbers and kinds of molecules found in a single cell at one specific point of time.

High-throughput sequencing technology is quickly becoming more affordable and as a result the forthcoming deluge of data poses unique challenges for bioinformatics, not only for handling all the data, but also for the completely new kinds of questions the biologists will want to ask from the data and the larger-scale experiments the new methods make possible, for example, sequencing samples from thousands of individual organisms or sequential samples from evolving microbial populations and so on.

2.3.2 *Proteomics data and degradomics*

Protein identification from small biological samples is advancing rapidly, so that also proteomics datasets are increasing exponentially. The most recent data analysis equipment are sensitive enough and the final bottlenecks of analysing the rich spectrometry data appears to have been solved (Cox & Mann 2008), so that ultra-high throughput and accurate protein determination is becoming possible and could generate similar amounts of data as the modern DNA sequencing machines. Fast protein identification then allows also manufacturing and use of specific protein expression analysis systems, if not already possible by abundance data from the mass spectra themselves. Traditional proteomics chips will though be continued to be used for quite a while still, because state of art spectrometric facilities are still quite expensive. Another proteomics database containing mass spectrometry data linked to genomic databases is MAPU database, which displays proteomic data from organelles, tissues and body fluids (Gnad *et al*. 2009).

Degradomics refers to the sets of small protein fragments resulting from the action of various protease enzymes within living cells. Modern mass spectrometric methods can now find the small degraded fragments of proteins among the other intact proteins in cell extracts, and the activities of specific proteases can be detected in this way (Schilling & Overall 2008), which is a requisite for protease systems biology. The Degradome database contains a listing of over 2,000 mammalian proteases, data about protein degradation products and diseases associated with proteolysis (Quesada *et al.* 2009).

2.3.3 *Protein expression profiles, 2D gel and protein interaction data*

Similarly to DNA expression microarray results standard MIAME, a proteomics MIAPE standard has been published (Taylor *et al.* 2007) to ensure compatibility and comparability of data from different laboratories. Large proteomics datasets are only recently being collected to integrated databases, including EBI-hosted PRIDE, which contains some 2,700 large-scale proteomics experiments with identifications of the proteins and methods of analysis (Jones *et al.* 2008). There is a multitude of proteomics experimental databases for different specific domains, so that the data accessibility situation is similar to the early stages of DNA expression databases, which have now been standardized to a larger extent than the protein expression databases.

Protein identification has traditionally been done by 2D gel electrophoresis producing protein spot data, in which proteins can be identified by their location and/or partial sequencing of isolated protein samples from the excised purified protein spots. The gel image data and annotation has been standardized by SwissProt to produce the SWISS-2DPAGE database (Hoogland *et al.* 2004), which in 2008 contained about 1,200 protein spot gels. SwissProt EXPASY website also contains listing and links to hundreds of specialized 2D protein spot databases. It is likely that new high-throughput technologies (*e.g.* mass spectrometry) will reduce the importance of gel image databases.

Protein interaction data are also slowly being integrated to centralized databases, though progress is still behind the DNA function and interaction databases. Notable protein-protein interaction databases include the HPRD, Human Proteins Reference Database (Prasad *et al.* 2008) containing some 38,000 interactions between about 25,000 proteins and Molecular INTeractions database MINT which covers proteins in all organisms and contains over 110,000 interactions between 30,000 proteins mined from published experimental data. IntAct database (Kerrien *et al.* 2008) at EBI contains 125,000 interactions for 55,000 proteins.

A new international collaboration has produced ProteinPedia (Kandasamy *et al.* 2009), which includes the HPRD as well as additional experimental data submissions from many laboratories around the world. Currently it contains some 2 million peptides and 5 million mass spectra and 35,000 interactions. It aims to be the definitive source of experimentally verified human protein interactions and is expanding rapidly. Currently over 50 terabytes of data is stored in 16 servers around the world using the Tranche file-sharing network at the consortium website http://www.proteomecommons.org/dev/dfs. The annotation includes post-translational protein modifications (*e.g.* phosphorylation of specific amino acids in the protein), tissue expression data, subcellular location, enzyme substrates and so on. Thus this resource is a significant advance in integration of protein related experimental data linked to other computational data about the known proteins.

The reliability of evidence and statistical significance cut-offs affect crucially the number of interactions to be taken into account in any global analysis. It is good to be aware of the types of experimental and *in silico* errors of evidence biasing specific interactions. Caution is needed when selecting the cut-offs for subsets of data to make sure that the quality of data is commensurate with the conclusions of analyses.

2.3.4 *Metabolomics and metabolic pathway databases*

Metabolomics databases contain data about non-proteinaceous and non-DNA compounds in living cells. These data includes small size carbohydrates, sugars and simple and complex organic chemicals. Thus it belongs partially into the cheminformatics domain. Nuclear magnetic resonance (NMR) data is an important source of such data. An important repository of NMR data is

Biological Magnetic Resonance Data Bank BMRB (http://www.bmrb.wisc.edu/), which includes a large number or small molecule related NMR spectra.

For the human metabolome of small molecules, the HMDB (www.hmdb.ca) is an important source of metabolomic information and contains a curated collection of human metabolite structures and functions as well as experimental data about concentrations of various small molecules (Wishart *et al.* 2009). The total number of molecules in version 2 is about 6,500, and for each item more than 100 chemical, biochemical and clinical data fields are included as well as links to protein databases and metabolic pathway databases. A more extensive database covering all organisms is the Madison Metabolomics Consortium Database (MMCD, Cui *et al.* 2008), which has some 20,000 compounds, including a large number of compounds from the plant *Arabidopsis*.

At the next level of annotation, metabolic compound and enzyme databases have been compiled to create metabolic pathway databases, which show the networks of biochemical processing of small molecules and proteins in living cells. Such databases are also abundant on the internet, with a multitude of small domain-specific databases dedicated to single organisms or specific metabolic pathways. A prominent generic global metabolic pathway databases is the Kyoto Encyclopedia of Genes and Genomes (KEGG), which is currently the premier freely accessible pathways database integrating genes and proteins, available at www.genome.jp/kegg. It currently contains data from 700 sequenced genomes, the related genes (orthologs) clustered to groups, and mapped to a global metabolism map called KEGG ATLAS (Okuda *et al.* 2008). KEGG ATLAS integrates over 200 metabolic maps in the KEGG system and organism-specific views of the maps are also available. Web-based display system iPath to browse and compare data interactively for selected species has also been developed (Letunic *et al.* 2008).

Other notable metabolic pathway databases include MetaCyc (Caspi *et al.* 2008) which contains some 1,500 pathways from 1,100 organisms (mostly bacteria and some plants), and a recently introduced MetaCrop, which contains pathways for commercially important crop plants rice, maize, wheat, barley, potato and canola (Grafahrend-Belau *et al.* 2008).

2.3.5 *Personalized data*

As high-throughput data acquisition methods continue to improve, more and more genetic data (DNA and protein sequences) and metabolomics and other biochemical data pertaining to individual people and individuals of other organisms is accumulating at increasing pace. These data include single nucleotide polymorphism (SNP) data, specific gene mutation data and metabolic marker data.

So far only two personal human genomes have been sequenced in full and carefully assembled, namely those of Dr James Watson and Dr. Craig Venter, both of human genome project fame. However, in 2010 the cost of sequencing a personal genome is expected to fall under US$ 10,000, and there is already a Personal Genome Project (http://www.personalgenomes.org/) which ultimately aims to sequence genomes of 100,000 individuals, first starting though only with the protein coding regions of the genome, which is about 1.5% of the whole genome length. Already data for the first set of 10 people is available, together with complete medical records. The goal of reducing the cost of an individual human genome below USD 1,000 is not too far off in the future.

2.3.5.1 *SNP data*

Quickly increasing amounts of partial or full genome sequences of individuals of the same species are becoming available, so that the concept of a "standard" genome sequence is becoming more nebulous. Even if a large number of individuals had been sequenced for their genome sequences, it is not clear that a consensus sequence of a "normal" or ancestral sequence could be constructed. This is due to both the presence of single nucleotide polymorphisms as well as chromosomal rearrangements at various length scales.

One important class of conserved segments of genome sequences is the haplotype variant, which encompasses segments of the genome which have consistent single nucleotide polymorphisms (SNPs) between different individuals. The conserved segments called haplotype blocks contain a conserved set of several SNPs. The haplotypes in the human genome have an average size of 18 kilobases (Matsuzaki *et al.* 2004).

Most SNPs have been determined from short sequencing reads, so that information about long-range covariation in the SNPs is not yet available. This is likely to change soon, because new methods are being developed, which utilize isolation of long DNA segments associated with specific SNP locations followed by high throughput sequencing and alignment of the long DNA segments (Dapprich *et al.* 2008). This will create an even more urgent need to develop methods to analyse conserved segments to be compared among large numbers of sequenced individuals.

The databases listing single nucleotide polymorphisms (SNPs) are already massive. The large human genome sequencing institutions have already accumulated huge catalogues of single nucleotide variations along the human genome. The current SNP database of the international HAPMAP project in November 2008 contains over 26 million SNPs from 11 different human populations. The NCBI SNP public depository contains over 50 million SNPs. The basic data for a typical entry from the NCBI SNP database is shown below.

```
rs45430330 [Plasmodium falciparum]
TTGGTGTGCTTCATCATTTTGGTGTG[C/T]CCCACCATTTT
GGTGTGTCCCACCA
```

The accession number rs45430330 is followed by the species name (here malaria parasite *P. falciparum*), and the sequence flanking the polymorphism is shown. The polymorphism indicated in the middle of the sequence is the nucleotide T instead of C in the "normal" standard sequence. Additional information in the database can include coordinates to the position of the full genome sequence or partial gene sequence.

Current SNP (single nucleotide polymorphism) analysis to compare different individuals currently relies on sequenced variation in only about 0.1% of the human genome obtained by microarray data focusing on the previously sequenced specific small regions of the genome, mostly protein-coding genes and their promoters. This will be soon replaced by large-scale ultra-high throughput complete sequencing of the whole genome from different individual organisms.

SNPs in pathogens will also be able to be discovered by real-time monitoring of bacterial and viral populations in diseased animals or humans by brute force sequencing of large numbers of individual bacterial and viral genomes during the progression and spread of the disease in an individual or in populations, *e.g.* HIV virus mutations during antiviral drug therapy. This is obviously of tremendous benefit in the research on pathogen evolution and methods to combat them to minimize dangerous mutant variants and to develop effective therapies.

2.3.5.2 *E-health data*

The future of individualized medicine relies on digital data, and is said to be essential to the envisaged P4 Medicine (Predictive, Preventive, Personalized, and Participatory). This still demands a major effort to harmonize the various data formats and semantic views of medical information, in addition to the challenges in actually obtaining the data and permissions to use it. The multitudinous formats and standards of archived national and institutional medical data is a formidable obstacle in this work. Several major existing databases are however serving as a good starting point.

One such database for human diseases is the OMIM, Online Mendelian Inheritance in Man, hosted by NCBI. It contains information on all known Mendelian inherited diseases and disorders and includes over 12,000 genes and known mutations for each disease. There are many other disease specific mutation databases, too. One example is a recent database of EGFR (epidermal growth factor receptor gene) somatic mutations (at http://www.somaticmutations-egfr.org/) containing over 3,300 mutations in a single gene causing cancer.

Most of these databases do not, however, include combined genetic and health data on specific individuals, but such databases are in construction in several countries. Many old patient health record databases are being complemented with genetic and sequencing data to achieve more integrated databases of increased value, together with national large biobanks to store DNA and other tissue samples with associated patient health data in large centres. The main current biobanks are listed in Table 2-4 below.

For the knowledge engineer who wants to use large-scale datasets of anonymised patient sample data for data mining, the procedures of getting access to data vary between projects, and there is no centralized resource. In the machine learning community, there is an excellent resource in the UCI Machine Learning Repository at http://archive.ics.uci.edu/ml/datasets.html, which contains some 48 life sciences datasets in a standardised format, but with only half a dozen datasets relating to DNA or protein sequences. The UCI collection is most suitable for benchmarking bioinformatics algorithms.

Table 2-4. Examples of current biobanks in 2008.

Databank	Samples	URL
Utah Population	6,500,000	http://www.hci.utah.edu/groups/ppr/
DeCode Biobank	110,000	http://www.decode.com/
UK Biobank	175,000	http://www.ukbiobank.ac.uk/
Finland Biobank	200,000	http://www.nationalbiobanks.fi/
Estonian Biobank	22,000	http://www.geenivaramu.ee/
German Kora-gen	18,000	http://epi.gsf.de/kora-gen/
LifeGene Sweden	500,000	http://www.lifegene.org/index_en.html

Thus there is a need for large standardized access to health databases with anonymised patient data, which could be used as test cases in bioinformatics algorithm comparisons and development. The datasets should be available also in a format that is easy to import to *e.g.* MatLab, which is one of the most commonly used platforms for artificial intelligence researchers. Genomics and sequence data commonly have their own differing data formatting standards, so that further harmonization would be very desirable. Many databases are already routinely available in XML format or in a format directly usable in BioPerl, a very common programming language in bioinformatics. Metadata standards are essential here, and such data is discussed next.

2.4 Semantic and processed text data

Large text datasets in general are difficult to visualize. For example, the Amazon bookshop has to be searched by keywords and by topic similarity searches or browsing of topic categories. The same applies to biology related scientific article databases like PubMed, Google Scholar contained publications, *etc.* PubMed abstracts of specific domains have been clustered to networks, which are available as "related documents" links in the PubMed website. Efficient visualization tools for browsing such large text databases are not yet available, but new good methods for clustering and visualization text data for PubMed abstract database are emerging (*e.g.* Theodosiu *et al.* 2008, which utilizes MCL algorithm).

Recently, the new field of biosemantics to address this problem is starting to gather momentum. The aim is to combine all biological knowledge to an integrated queriable system. This demands that concepts of biological entities, their functions and the relationships between entities and functions can be inferred automatically from semantically parsed text data, including even new concepts and classifications that ensue due to new information obtained from newly accumulated novel text data of scientific publications.

A recent archival database in the website http://bionlp-corpora.sourceforge.net/index.shtml stores a large collection of biology related text corpora, which include protein interactions, gene homonyms, Medline mined data, human annotated biological articles text and human produced GeneRIFs. GeneRIFs are short 255 characters long descriptions of gene function produced by NCBI indexing staff to annotate submitted gene sequences (Entrez database entries) with their biological function data. Two good recent books on biological text mining are Ananiadou & McNaught (2006) and Shatkay & Craven (2007).

2.4.1 *Ontologies*

Semantic web technologies for bioinformatics are merely starting. Central to the development of metadata for knowledge representation are standardized ontologies. The initial step has been the construction of domain-specific ontologies, which then form the basis of controlled vocabularies and machine-readable concepts. Ontologies can then be used by software agents for searching, retrieving and analyzing information for a client (which could be a software agent as well). Large ontologies with attached controlled vocabularies are databases in their own right even before the insertion of the raw data pertaining to the classes and subclasses representing the concepts in the ontology.

2.4.1.1 *Bioinformatics ontologies*

In addition to the already mentioned pioneering Gene Ontology, recently a multitude of subdomain ontologies has emerged. At present, over 60 groups participate under the OBO Foundry umbrella (http://www.obofoundry.org), with the objective of developing interoperable ontologies for the whole of biological research (Smith *et al.* 2007). The OBO Foundry database lists a large number of biological ontologies, including the pioneering

Gene Ontology (GO), Cell Ontology, Foundational Model of Anatomy, Chemical Entities of Biological Interest, Disease Ontology and so on. Mapping and linking of the ontologies to each other is also already available for several pairs of ontologies. An interesting taxonomic ontology has also been included, which could subsume NCBI taxonomy data (Schulz *et al.* 2008).

In addition to this central repository of biological ontologies, there are numerous separate efforts for semantic standardization of various domains of biology and biodata annotation, many of which still await standardization for compatibility with each other and the OBO Foundry. A recent example is the Protein Feature Ontology, which tries to standardize the annotation of protein features (Reeves *et al.* 2008).

2.4.1.2 *Medical ontologies*

Medical ontologies have a long historical background, but a full review of the many already existing medical ontologies is beyond the scope of this book. A good overview of the current situation is available in Lussier & Bodenreider (2007).

One of the most important is the Unified Medical Language System (UMLS), started in 1986, which is a huge database of controlled vocabularies in the biomedical sciences (Bodenreider 2004). The core MetaThesaurus combines over 100 vocabularies of different fields, with 1 million medical concepts and 5 million concept names. The vocabularies are still only loosely connected, so that the UMLS as a whole cannot be said to be a complete semantically integrated medical ontology.

UMLS includes NCBI Taxonomy, Gene Ontology, OMIM, SNOMED (database of clinical symptoms and diagnosis concepts) and MESH, among others. MESH (Medical Subject Headings) is an important vocabulary thesaurus which contains some 24,000 main indexing terms. MESH was developed by the U.S. National Library of Medicine (NLM), which uses it for indexing the biomedical literature in the MEDLINE database, which currently in 2009 comprises more than 19 million biomedical article references and abstracts. Each MEDLINE article is tagged with about a dozen keywords representing the subject matter of the article. This database is integrated into the PubMed abstract database and enables the linking of abstract to "Related documents" at NCBI.

MESH is important, because it can be reused for linking to a variety of other data. For example, CleanEx database of gene expression datasets has been indexed by adding MESH annotation tag to each expressed gene in the included microarray experiments (Praz & Bucher 2009), so that the microarray experiment repository can be searched more easily with MESH terms, which relate to diseases, genes, protein functions etc.

Many of such controlled vocabularies have been built independently with differing formats and standards, and do not always conform to good ontological standards. It is generally now recognized, that the maintenance and quality control of big concept databases is challenging, especially for multidomain medical ontologies (Rogers 2006). Many inconsistencies and mistakes remain to be assessed, as shown, for example, in a case study of SNOMED (Ceusters *et al.* 2003).

2.4.2 *Text-mined annotation data*

The natural language processing community has started to make an impact in bioinformatics also, enabling annotation systems to link and cluster related text documents based on commonly co-occurring biological terms and gene and protein names and so on. Such methods are still quite compute intensive, because they require the parsing of the text to semantic units and the selection of informative words and concepts for clustering. Furthermore, the extraction of high-quality semantic relations is difficult.

A pilot study using full text scientific documents about yeast metabolism was mined for gene functions, but a disappointingly large numbers of false positives and false negatives were found (Crangle *et al.* 2007); precision (ratio of correct answers to total answers) was only 0.92%. This result is likely to hold for other non-trivial biological text mining tasks as well, due to the huge complexity of biology as a discipline. Thus it seems likely, that solid semantic data extraction systems for biological publications are far in the future.

However, for specialized narrow tasks, text mining approaches can be successful; for example, a protein phosphorylation database was enhanced by mining text of PubMed abstracts for phosphorylation substrates and predicted phosphorylation sites with a very good precision of 91.4% (Hu *et al.* 2005).

2.5 Integrated and federated databases

In one aspect, the global major sequence databanks in NCBI and EBI are already integrated databases, because a lot of annotations and hyperlinks to related information are available. This is best suited, however, to human browsing and follow-up of the information to relevant linked information in other websites and databases.

For semantic web technologies to have an impact, one needs the following building blocks and infrastructure to be set up in a way that they work together in a B2B type of environment to achieve automated semantic search and analysis of huge biological datasets (Shah *et al.* 2008):

Web based methods for:
- Semantic Web for machine-readable ontologies and concepts
- Web Services (Net 2.0) for standardized information transfer
- Webified software agents for search and analysis requests
Grid-based methods for:
- Distributed data distribution and work unit allocation
- Distributed computational processing
- Distributed results retrieval and merging at final destination

Automated methods are important in using and integrating data from different sources and databases in order to evaluate how compatible two databases are semantically and whether they can be automatically integrated in web and grid services. This aspect of web grid technology is just beginning to be developed (see *e.g.* Chen *et al.* 2006). For semantic data to be detailed enough and for the ontology to appear as a true network of bidirectional relationships, it appears that XML data format is not sufficient, and use of RDF and OWL technologies are needed, with the use of location-independent URI (Universal Resource Identifiers).

Web services have already been built for some bioinformatics data domains. A list of biological data services based on web/grid technologies is available at the MyGrid consortium at http://www.mygrid.org.uk/wiki/Mygrid/BiologicalWebServices. The list includes many major databanks in biology offering web grid standardized views of web services. Description of the EBI web services is given in Labarga *et al.* (2007).

Ultimately agent-based Web-services aim to provide autonomously acting software that communicates with other agents and services using semantic messaging in a high-level semantic language. This is envisaged to happen in a service-oriented semantic grid of distributed computing resources and data. Ideally, for ease of use for the non-expert users, such a grid based computing service should be a ready-to-use web-based application server linking the offered data to standard analysis methods on the grid. Such services could include BLAST, phylogenetic tree calculations and other CPU-intensive algorithms.

The European Union Enabling Grids for E-science in Europe (EGEE) has already made available over 20 grid applications, including the consortia BioInfoGrid (http://www.bioinfogrid.eu/), WISDOM (http://wisdom.eu-egee.fr) and EMBRACE (http://www.embracegrid.info).

Currently Globus-based grids are common, but have the disadvantage that they are centralized, requiring a central meta-scheduler and master/slave grids are suitable mostly for coarse-grained jobs. Ideally, the grid systems would be dynamic, or organic grid systems (Chakravarti *et al.* 2006), so that the software agent dispatched from the user would independently find the best node to do the job, carrying the execution code and scheduling data with it. This would lead to de-centralized dynamic scheduling and would work in a grid with large numbers of nodes and flexible peer-to-peer handling of jobs. The agents should be strongly mobile ones controlling their own behaviour.

Thus it seems that there are still many technological limitations in the envisaged computational grid adapted semantic query technologies, mainly due to the limited scope and interoperability of various biology domain ontologies, and limits in the expressivity of ontology representations to match complexities of biology data. Also, the move from the limitations of XML to the more flexible RDF methods is not yet sufficiently progressed. The Directed Acyclic Graph (DAG) representation of some biological ontologies (notably GeneOntology) is also not sufficiently expressive for formal logic processing of ontological analysis and decision-making. Therefore the grid computing systems for biology are still deemed to be semi-mature (Shah *et al.* 2008), requiring better links between application and middleware layers and easy methods to plug-in desired bioinformatics algorithms.

There is also the problem of the different database ontologies not matching each other, so that there needs to be several mappings of data source formats to requester's ontology. This could be achieved using the Web Services Management Language (WSML). The main task is to produce the formal ontologies which are sufficiently matching to enable proper communication between client nodes and data/computing service provider nodes in the true B2B communication scenario.

In the biomedical information retrieval area, the first prototype of a meta-service (called BioCreative MetaServer system) for biomedical information extraction has been built (Leitner *et al.* 2008), and contains federated databases in several countries, based on a set of 28,000 PubMed abstracts which have been annotated for gene/protein names linked to sequence databases, taxon classifications and protein-protein interactions. Data is available via web services or via programmatic XML-RPC queries. This is a promising proof of principle service showing the way forward.

References

Ananiadou, S. & McNaught, J. 2006. Text mining for biology and medicine. *Boston/London: Artech House.*

Andronescu, M., Bereg, V., Hoos, H.H. & Condon, A. 2008. RNA STRAND: The RNA secondary structure and statistical analysis database. *BMC Bioinformatics 9(1): 340.*

Anonymous 2008a. http://www.genomesonline.org/gold.cgi?want=Metagenomes

Anonymous 2008b. http://www.metahit.eu/metahit/

Anonymous 2008c. http://www.ensembl.org

Anonymous 2008d. http://www.wwPDB.org

Anonymous 2008e. http://www.ccdc.cam.ac.uk/products/csd/

Anonymous 2008f. http://www.epigenome.org

Attwood, T.K., Bradley, P., Flower, D.R., Gaulton, A, Maudling, N, et al. 2003. PRINTS and its automatic supplement, prePRINTS. *Nucleic Acids Research 31: 400–402.*

Barrell, D., Dimmer, E., Huntley, R. P., Binns, D., O'Donovan, C. et al. 2009. The GOA database in 2009—an integrated Gene Ontology Annotation resource *Nucleic Acids Research 37: D396–D403.*

Barrett, T., Troup, D.B., Wilhite, S.E., Ledoux, P., Rudnev, D. et al. 2009. NCBI GEO: archive for high-throughput functional genomic data. *Nucleic Acids Research 37: D885–890.*

Basu, S., Bremer, E., Zhou, C. & Bogenhagen, D.F. 2006. MiGenes: a searchable interspecies database of mitochondrial proteins curated using gene ontology annotation. *Bioinformatics 22(4): 485–492.*

Bateman, A., Letunic, I., Doerks, T. & Bork, P. 2008. SMART 6: recent updates and new developments *Nucleic Acids Research 37: D229–D232.*

Bernatavichute, Y.V., Zhang, X., Cokus, S., Pellegrini, M. & Jacobsen, S.E. 2008. Genome-wide association of histone H3 lysine nine methylation with CHG DNA methylation in *Arabidopsis thaliana. PLoS ONE 3(9): e3156.*

Bock, C., Walter, J., Paulsen, M. & Lengauer, T. 2008. Inter-individual variation of DNA methylation and its implications for large-scale epigenome mapping. *Nucleic Acids Research 36(10): 55–66.*

Bodenreider, O. 2004. The Unified Medical Language System (UMLS): integrating biomedical terminology. *Nucleic Acids Research 32: D267–270.*

Caspi, R., Foerster, H., Fulcher, C.A., Kaipa, P., Krummenacker, M., et al. 2008. The MetaCyc Database of metabolic pathways and enzymes and the BioCyc collection of Pathway/Genome Databases. *Nucleic Acids Res. 36: D623–631.*

Chakravarti, A.J., Baumgartner, G. & Lauria, M. 2005. The Organic Grid: selforganizing computational biology on desktop grids, In: Zomaya, A.Y. (Ed), Parallel Computing for Bioinformatics and Computational Biology: Models, Enabling Technologies and Case Studies, John Wiley & Sons, pp. 671–703.

Chatraryamontri, A., Ceol, A., Palazzi, L.M., Nardelli, G. et al. 2006. MINT: the Molecular INTeraction database. *Nucleic Acids Research 35: D572–D574.*

Chen, L., Shadbolt, N.R., Goble, C. & Tao, F. 2006. Managing Semantic Metadata for Web/Grid Services. *International Journal Web Services Research 3(4): 73–94.*

Cokus, S.J., Feng, S., Zhang, X., Chen, Z., Merriman *et al.* 2008. Shotgun bisulphite sequencing of the Arabidopsis genome reveals DNA methylation patterning. *Nature 452: 215–219.*

Cox, J., Mann, M. 2008. MaxQuant enables high peptide identification rates, individualized p.p.b.-range mass accuracies and proteome-wide protein quantification. *Nature Biotechnology 26(12): 1367–1372.*

Cui, Q., Lewis, I.A., Hegeman, A.D., Anderson, M.E., Li, J. et al. 2008. Metabolite identification via the Madison Metabolomics Consortium Database. *Nature Biotechnology 26: 162–164.*

Dapprich, J., Ferriola, D., Magira, E.E., Kunkel, M. & Monos, D. 2008. SNP-specific extraction of haplotype-resolved targeted genomic regions. *Nucleic Acids Research e94: 1–9.*

Dinger, M.E., Pang, K.C., Mercer, T.R., Crowe, M.L., Grimmond, S.M. & Mattick, J.S. 2009. NRED: a database of long noncoding RNA expression. *Nucleic Acids Research 37: D122–D126.*

Field, D., Garrity, G.M., Gray, T., Morrison, N., Selengut, J. et al. 2008. Towards a richer description of our complete collection of genomes and metagenomes: the "Minimum Information about a Genome Sequence" (MIGS) specification. *Nature Biotechnology 26: 541–547.*

Figueroa, M.E., Reimers, M., Thompson, R.F., Ye, K., Li, Y. *et al.* 2008. An integrative genomic and epigenomic approach for the study of transcriptional regulation. *PLoS ONE 3(3): e1882.*

Garrity, G.M., Field, D., Kyrpides, N, Hirschman, L., Sansone, S.-A. et al. 2008. Toward a Standards-Compliant Genomic and Metagenomic Publication Record. *OMICS, A Journal of Integrative Biology, 12(2): 157–160.*

Gnad, F., Oroshi, M., Birney, E. & Mann, M. 2009. MAPU 2.0: high-accuracy proteomes mapped to genomes. *Nucleic Acids Res. 37(Database issue): D902–906.*

Godfray, C., Clark, B.R., Kitching, I.J., Mayo, S.J. & Scoble, M.J. 2007. The web and the structure of taxonomy. *Systematic Biology 56: 943–955.*

Grafahrend-Belau, E., Weise, S., Koschützki, D., Scholz, U., Junker, B.H. & and Schreiber, F. 2008. MetaCrop: a detailed database of crop plant metabolism. *Nucleic Acids Research 36: D954–D958.*

Griffiths-Jones, S., Moxon, S., Marshall, M., Khanna, A., Eddy, S.R. & Bateman, A. 2005. Rfam: annotating non-coding RNAs in complete genomes. *Nucleic Acids Research 33: D121–D124.*

Henikoff, J.G., Greene, E.A., Pietrokovski, S. & Henikoff, S. 2000. Increased coverage of protein families with the BLOCKS database servers. *Nucleic Acids Research 28: 228–230.*

Henrick, K., Feng, Z., Bluhm, W.G., Dimitropoulos, D., Doreleijers, J.F. et al. 2008. Remediation of the protein data bank archive. *Nucleic Acids Research 36 Database issue: D426–D433.*

Hoogland, C., Mostaguir, K., Sanchez, J.-C., Hochstrasser, D.F. & Appel, R.D. 2004. SWISS-2DPAGE, ten years later. *Proteomics 4(8): 2352–2356.*

Hu, Z.Z., Narayanaswamy, M., Ravikumar, K.E., Vijay-Shanker, K., & Wu, C.H. 2005. Literature mining and database annotation of protein phosphorylation using a rule-based system. *Bioinformatics 21(11): 2759–2765.*

Hunter, S., Apweiler, R., Attwood, T.K., Bairoch, A. et al. 2008. InterPro: the integrative protein signature database. *Nucleic Acids Research 37: D211–D215.*

Jiang Z, Rokhsar DS, Harland RM. 2009. Old can be new again: HAPPY whole genome sequencing, mapping and assembly. *Int J Biol Sci. 5(4): 298–303.*

Jones, P., Côté, R.G., Cho, S.Y., Klie, S., Martens, L. et al. 2008. PRIDE: new developments and new datasets. *Nucleic Acids Research 36: D878–D883.*

Juncker, A.S., Jensen, L.J., Pierleoni, A., Bernsel, A., Tress, M.L., Bork, P., von Heijne, G., Valencia, A., Ouzounis, C.A., Casadio, R. & Brunak, S. 2009. Sequence-based feature prediction and annotation of proteins. *Genome Biol. 10(2): 206.*

Kahvejian, A., Quackenbush, J., Thompson, J.F. 2008. What would you do if you could sequence everything? *Nature Biotechnology 26(10): 1125–1133.*

Kandasamy, K., Keerthikumar, S., Goel, R., Mathivanan, S., Patankar, N. et al. 2009. Human Proteinpedia: a unified discovery resource for proteomics research. *Nucleic Acids Research 37: D773–781.*

Kerrien, S., Alam-Faruque, Y., Aranda, B., Bancarz, I., Bridge, A. et al. 2007. IntAct: open source resource for molecular interaction data. *Nucleic Acids Research 35: D561–D565.*

Klimke, W., Agarwala, R., Badretdin, A., Chetvernin, S., Ciufo, S. et al. 2009. The National Center for Biotechnology Information's Protein Clusters Database. *Nucleic Acids Research 37: D216–D223.*

Labarga, A ., Valentin, F., Andersson, M. & Lopez, R. 2007. Web Services at the European Bioinformatics Institute. Nucleic Acids Research 35: W6–11.

Leitner, F., Krallinger, M., Rodriguez-Penagos, C., Hakenberg, J. & Plake, C. 2008. Introducing meta-services for biomedical information extraction. *Genome Biology 9(Suppl 2): S6.*

Letunic, I., Yamada, T., Kanehisa, M. & Bork, P. 2008. iPath: interactive exploration of biochemical pathways and networks. *Trends in Biochemical Sciences 33: 101–103.*

Liolios, K., Mavrommatis, K., Tavernarakis, N. & Kyrpides, N.C. 2008. The Genomes On Line Database (GOLD) in 2007: status of genomic and metagenomic projects and their associated metadata. *Nucleic Acids Research 36: D475–D479.*

Lussier, Y.A. & Bodenreider, O. 2007. Clinical Ontologies for Discovery Applications. In: Semantic Web: Revolutionizing Knowledge Discovery in the Life Sciences, C.J.O. Baker & K-H. Cheung (Eds.), Springer.

Marchler-Bauer, A., Anderson, J.B., Chitsaz, F., Derbyshire, M.K., DeWeese-Scott, C. et al. 2009. CDD: specific functional annotation with the Conserved Domain Database. *Nucleic Acids Research 37: D205–210.*

Matsuzaki, H., Dong, S., Loi, H., Di, X., Liu, G. et al. 2004. Genotyping over 100,000 SNPs on a pair of oligonucleotide arrays. *Nature Methods 1: 109–111.*

Mulder, N.J., Kersey, P., Pruess, M. & Apweiler, R. 2008. In silico characterization of proteins: UniProt, InterPro and Integr8. *Molecular Biotechnology 38(2): 165–177.*

Okuda, S., Yamada, T., Hamajima, M., Itoh, M., Katayama, T. et al. 2008. KEGG Atlas mapping for global analysis of metabolic pathways. *Nucleic Acids Research 36: W423–W426.*

Pachkov, M., Erb, I., Molina, N. & van Nimwegen, E. 2007. SwissRegulon: a database of genome-wide annotations of regulatory sites. *Nucleic Acids Research 35: D127–D131.*

Parkinson, H., Kapushesky, M., Kolesnikov, N., Rustici, G., Shojatalab, M. *et al.* 2008. ArrayExpress update—from an archive of functional genomics experiments to the atlas of gene expression. *Nucleic Acids Research 37 (Supplement 1): D868.*

Prasad, T. S. K., Kumaran, G., Kandasamy, K., Keerthikumar, S., Kumar, S., Mathivanan, S. et al. 2008. Human Protein Reference Database—2009 Update. *Nucleic Acids Research 37: D767–D772.*

Praz, V. & Bucher, P. 2009. CleanEx: new data extraction and merging tools based on MeSH term annotation. *Nucleic Acids Research 37: D880–D884.*

Quesada, V., Ordóñez, G.R., Sánchez, L.M., Puente, X.S., López-Otín, C. 2009. The Degradome database: mammalian proteases and diseases of proteolysis. *Nucleic Acids Res. 37: D239–43.*

Reeves, G.A., Eilbeck, K., Magrane, M., O'Donovan, C., Montecchi-Palazzi, L., Harris, M.A., Orchard, S., Jimenez, R.C., Prlic, A., Hubbard, T.J.P. et al. 2008. The Protein Feature Ontology: a tool for the unification of protein feature annotations *Bioinformatics 24(23): 2767–2772.*

Rogers, J.E. 2006. Quality assurance of medical ontologies. *Methods of Informatics in Medicine 45(3): 267–274.*

Sandelin, A., Alkema, W., Engström, P., Wasserman, W.W. & Lenhard, B. 2004. JASPAR: an open-access database for eukaryotic transcription factor binding profiles. *Nucleic Acids Research 32: D91–D94.*

Schilling, O. & Overall, C.M. 2008. Proteome-derived, database-searchable peptide libraries for identifying protease cleavage sites. *Nature Biotechnology 26(6): 652–653.*

Schulz, S., Stenzhorn, H. & Boeker, M. 2008. The ontology of biological taxa. *Bioinformatics 24: i313–i321.*

Schumacher, A., Wang, S.C., Petronis, A., Mill, J., Tang, T., Kaminsky, Z., Khare, T., Yazdanpanah, S., Bouchard, L., Jia, P., Assadzadeh, A. & Flanagan, J. 2008. Epigenomic profiling reveals DNA-methylation changes associated with major psychosis. *American Journal of Human Genetics 82(3): 696–711.*

Shah, A.A., Barthel, D., Lukasiak, P., Blazewicz, J. & Krasnogor, N. 2008. Web & Grid Technologies in Bioinformatics, Computational and Systems Biology: A Review. - *Current Bioinformatics 3(1): 10–31.*

Shatkay, H. & Craven, M. 2007. Biomedical text mining. *Cambridge (Massachusetts): MIT Press.*

Sigrist C.J.A., Cerutti L., Hulo N., Gattiker A., Falquet L., Pagni M., Bairoch A. & Bucher P. 2002. PROSITE: a documented database using patterns and profiles as motif descriptors. *Briefings in Bioinformatics 3: 265–274.*

Sigrist, C.J.A., De Castro, E., Langendijk-Genevaux, P.S., Le Saux, V., Bairoch, A. & Hulo, N. 2005. ProRule: a new database containing functional and structural information on PROSITE profiles. *Bioinformatics 21: 4060–4066.*

Smith, B., Ashburner, M., Rosse, C., Bard, J., Bug, W., Ceusters, W. et al. (2007). The OBO foundry: coordinated evolution of ontologies to support biomedical data integration. *Nature Biotechnology 25: 1251–1255.*

Tatusov, R.L., Fedorova, N.D., Jackson, J.D., Jacobs, A.R., Kiryutin, B. et al. 2003. The COG database: an updated version includes eukaryotes. *BMC Bioinformatics 4: 41.*

Taylor, C.F., Paton, N.W., Lilley, K.S., Binz, P.A., Julian, R.K.Jr. et al. 2007. The minimum information about a proteomics experiment (MIAPE). *Nature Biotechnology 25(8): 887–893.*

Theodosiou, T., Darzentas, N., Angelis, L. & Ouzounis, C.A. 2008. PuReD-MCL: a graph-based PubMed document clustering methodology. *Bioinformatics 24(17): 1935–1941.*

Wishart, D.S., Knox, C., Guo, A.C., Eisner, R., Young, N. et al. 2009. HMDB: a knowledgebase for the human metabolome. *Nucleic Acids Research 37: D603–D610.*

Xie, X-Q. & Chen, J-Z. 2008. Data Mining a Small Molecule Drug Screening Representative Subset from NIH PubChem. *Journal of Chemical Informatics and Modelling 48(3): 465–475.*

Zhu, J. & Zhang, M.Q. 1999. SCPD: a promoter database of the yeast Saccharomyces cerevisiae. *Bioinformatics 15: 607–611.*

Zilberman, D. & Henikoff, S. 2007. Genome-wide analysis of DNA methylation patterns. *Development 134: 3959–3965.*

Zilberman, D. 2008. The evolving functions of DNA methylation. *Current Opinion in Plant Biology 11: 554–559.*

Chapter 3

Local Pattern Discovery and Comparing Genes and Proteins

Local patterns are sequence motifs that span only a portion of the DNA or protein sequence to be analyzed or compared, as opposed to global patterns, which pertain to the whole sequence. Local patterns constitute a local, compact representation of some essential features of the genes or proteins, and are especially useful for describing protein domains. Protein domains are short conserved segments of genes that are conserved and combined with different domains to create diversity in the function of evolving proteins. In addition, small local DNA and RNA motifs can correspond to the DNA binding domain of proteins interacting with the DNA (*e.g.* transcription factors). DNAs and RNAs can also have short domains which bind to other interacting DNAs or RNAs, for example microRNAs binding to the DNA or RNA.

Local pattern search algorithms are used to detect and utilize such short conserved motifs in the same gene family among different organisms. Short motifs are utilized by evolution to modify existing proteins by shuffling of DNA and thus the resulting corresponding short protein motifs in the translated proteins. In protein evolution, motifs or domains of about 40 amino acids are thought to be shuffled around to create new proteins with new functionalities. This is accomplished by moving the DNA exons (protein coding segments) of DNA around in the genome by translocations. The process is called mix and match mode in exon shuffling (Schmidt & Davies 2008, Moore *et al.* 2008). Such protein evolution can also involve duplication of domains, leading to complete or partial repeats of motifs in a protein (Bjorklund *et al.* 2006). In vertebrate proteomes, half of the protein domains reside in repeat regions of the proteins.

Local pattern searches can thus not only group together similar proteins with putatively similar functions, but also give clues to the evolutionary history of the protein family within and between species. Thus local patterns complement phylogenetic analysis by global sequence alignments, which will be covered in chapter 4.

Many algorithms have been developed for local motif searches; a good general review for DNA motif methods is in Das & Dai (2007) and general practical advice in MacIsaac & Fraenkel (2006). The methods discussed in this chapter are a representative sample of the most useful and widely used methods, but new methods keep appearing.

The basic task is to find conserved motifs which are statistically significant from the background occurrences of "random" motifs in the overall sequence or set of sequences. The first stage is to discover the motifs, then to reduce the redundancy of the discovered motifs to cluster them to consistent groupings and then optionally to relate them to previously known motifs or other annotations pertaining to the location of the motif in the sequence.

In general, most of the past motif algorithms were based on gapless motifs, that is, where the sequences vary only by substitutions, with no insertions or deletions of nucleotides or amino acids. These algorithms have been quite successful, but do not fully encompass the biological reality, because many DNA and protein motifs are of flexible length. The difficulty is identifying flexible size motifs with insertions and deletions in the combinatorial explosion of possibilities which need to be searched.

The algorithms for motif finding can be divided into two main classes: probabilistic methods based on profiles and word string (k-mer) based methods.

The probabilistic approach represents the motifs by a profile or Position Weight Matrix (PWM, also called PSSM for Position Specific Sequence Matrix), which gives the probabilities of each nucleotide or amino acid in each position in the aligned set of subsequences. Search through the dataset is not exhaustive, but probabilistic, making it a fast method for big datasets with a large number of sequences and/or long sequences. However, a true global maximum (*i.e.* the best motif) cannot be guaranteed. Both single-motif methods (subchapter 3.1.1) and multiple-motif methods (subchapter 3.1.2) have been developed.

The word or string based approaches evaluate and compare oligomer frequencies (k-mers) of the aligned subsequences. Clusters of similar k-mers are then built for multiple alignment and for making the consensus sequence. An exhaustive search is possible, provided the data set is not too large. Informative k-mers are a powerful approach for short consistent motifs (subchapter 3.1.3) and are also useful for identification of genomic fragments from metagenomic samples (see later subchapter 4.2.1).

As more than a hundred different methods have been published for local motif finding, there is an overabundance of algorithms. Among the large number of new recently published methods, it appears that the profile based methods are still most popular, especially the traditional Expectation maximum (EM) based MEME (see subchapter 3.1.1) which performs well in most applications. MEME also does well compared to Gibbs sampling based methods, though this may differ for specific datasets. Quest *et al.* (2008) recently developed a good methodology for benchmarking and comparison for motif finders.

In integrated motif annotation methods the comparison and merging of found motifs to known motifs can be achieved by similarity comparisons, often with criteria that are different from finding the *de novo* motifs. Usually more stringent criteria for accepting the new motifs are needed, because the known motifs are normally more reliable information than the *de novo* motifs obtained from a limited data set in the analysis of a novel sequence or group of sequences.

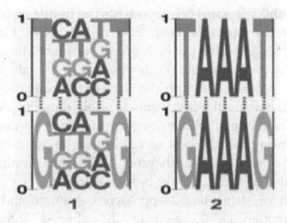

Figure 3-1. Two pairs of aligned motifs with three conserved positions in the middle. In the sequence logo format the height of the letter indicates the probability of the letter in the set of aligned sequences. The motif pair on left has less informative positions than the motif on right, and the similarity significance scoring should reflect this. Adapted from Habib *et al.* (2008). Reproduced by Creative Commons Attribution License from an Open Access article.

A technical note on how to assess similarity of motifs found by different methods is appropriate here (see Figure 3-1). Habib *et al.* (2008) pointed out that in aligning short motif profiles and scoring conserved positions, none of the currently commonly used scoring systems (Pearson correlation, Jensen-Shannon divergence, Euclidean distance *etc.*) actually performs well, due to the poor distinction of informative and non-informative positions, because an independence of each conserved position is assumed. Habib *et al.* developed a new Bayesian Likelihood 2-Component (BLiC) score, which weights the compared motif positions by their information content and showed that this scoring performed better than the traditional scoring systems for motif similarity. This shows that motif similarity estimation should be treated in a similar way to sequence multiple alignment, albeit at a local scale in ungapped motifs. It would be useful to extend of the BLiC scoring method to comparison of gapped motifs as well.

3.1 DNA/RNA motif discovery

The DNA alphabet has only 4 letters, A, T, C and G, which means that the information content per letter is quite small, compared to the amino acid alphabet with 21 amino acids. Therefore it is clear that for short DNA motifs much less information is available for classification algorithms than when analyzing protein motifs of similar length. This means that for short motifs, a larger training and validation set is also necessary. Also, because mutations in single bases occur easily, the biological variation in experimental data can mean that the input data is noisy. Protein sequences, in contrast, are evolutionarily much more stable, because many DNA mutations are silent and do not cause a change in the encoded protein. Therefore in DNA motif analysis, more latitude has to be given to the possibility of random similarities in the found sequences containing a putative sequence motif.

An important technical issue for DNA motif algorithms is the motif significance evaluation. Hundreds of different algorithms have been published, often relying on different motif significance metrics. Attempts to standardize motif statistics and evaluation have already started, see *e.g.* major comparison of the main motif finders in Tompa *et al.* (2005) and recent discussion in Sandve & Drabløs (2006), Wei & Yu (2005) and Klepper *et al.* (2008). A full survey of DNA motif methods is not attempted here, but the leading algorithms are introduced and examples of novel innovative approaches are also given in this section.

3.1.1 *Single motif models: MEME, AlignAce etc.*

Probabilistic profile-based approaches come in two main flavours, Gibbs sampling based, and Expectation Maximization based algorithms. Gibbs sampling based methods can be regarded as a Markov Chain Monte Carlo (MCMC) approach. In the first step, the Markov-Chain is used to select the next nucleotide or amino acid in the linear motif and each step depends only on the results of the preceding step. In the second step, the Monte-Carlo part consists of a probabilistic randomized sample of previous step candidates.

Expectation Maximization (EM) methods are also iterative optimization methods which estimate the model at each step and search for the best Position Weight Matrix to fit the multiple alignment. However, simple maximum likelihood estimation is normally not possible for typical motif finding problems where there are latent variables and/or missing data. Therefore an iterative two-step procedure is needed to build a probabilistic model for the motif.

In the EM methods the first expectation step (E-step) computes the probability distribution of motif letter positions for each sequence. The second maximization step (M-step) resets the motif letter frequencies based on the expected letter counts for each position in the motif. The maximization step can be regarded as maximization of the expected log-likelihood of the data. The maximization step drives the model toward a converging optimum by constrained guesses of parameter values, that is, solving the whole problem by stepwise local optimizations. A lucid explanation of the working of EM method is in Do & Batzoglou (2008).

EM methods do not guarantee reaching the global optimum, because they are only locally optimal; they are also sensitive to the initialization parameters. However, in general EM methods are robust and simple to implement, are guaranteed to converge to a stable solution and work well in practice in most ungapped motif discovery tasks.

MEME is one of the most popular Expectation Maximum based methods, and has proved to be very helpful in finding ungapped motifs in both DNA and proteins (Bailey *et al.* 2006). The web server is at http://meme.nbcr.net/meme4/intro.html. The output of the web service includes sequence logos, which show the information content of each aligned residue in the motif by the height of the residue letter. An example with protein sequences is discussed below. In DNA analysis by MEME, the only main difference is that more data (or a better conserved motif) is needed for statistical significance, due to less information content per letter in DNA (alphabet of 4) than in proteins (alphabet of 21). Extension of the MEME method to gapped motifs with arbitrary insertions and deletions has been recently proposed (Frith *et al.* 2008), leading to the development of the GLAM2 family of algorithms, which are also available on the MEME website.

Standard MEME outputs an ungapped multiple alignment of each motif and displays also a map of the locations of the motifs, as shown in Figure 3-2. In this example the three motifs are not in the same order in all sequences, but all three motifs are nevertheless easily seen to be present in all the input sequences, in some cases more than once.

It is instructive to compare the set of motifs discovered by MEME and GLAM2 from a small set of *Arabidopsis* plant sequences (example dataset from the MEME website). Figure 3-3 shows the three ungapped motifs found by MEME.

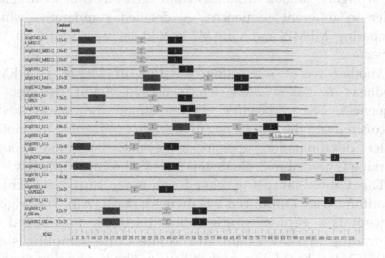

Figure 3-2. Motif summary output from MEME 4.0, showing the location of the found ungapped motifs. The webserver example output shows coloured motifs for easy visualization of the motif groupings.

GLAM2 results in Figure 3-4 show that very similar, but longer motifs are found. For the first motif, the highly conserved amino acids H, D, K and N (shown in bold as tall letters) are clearly identified by both MEME and GLAM2. However, GLAM2 identified another conserved area nearby to the right of the MEME motif, having the highly conserved residues D, F and G. Thus the gapped motif finding resulted in a longer and more informative motif than the ungapped motif method of MEME.

The multiple alignment of the first GLAM2 motif in Figure 3-5 also shows that the alignment includes added gaps, which improves the motif detection and motif quality. The improved quality motif with gaps would then be more efficient in searching databases for sequences with similar motifs, so that the number of false negatives would be lower. One possible disadvantage of using the longer motifs generated by GLAM2 is that they may merge two nearby motifs together, which might be better described by two independent motifs, especially if the two motifs are at highly variable distances from each other, or in different order in different sequences. A detailed comparative study comparing MEME and GLAM2 would be very useful in this respect.

Another recent interesting variation of MEME is DEME, (Redhead & Bailey 2007) which uses as input two sets of sequences, one a positive set containing the motifs to be discovered and the other a negative set, not containing the motifs. Better discrimination appears achievable, but more bench-marking is needed to evaluate this new method appropriately.

The Gibbs sampling based sequence motif searching algorithms were introduced by Lawrence *et al.* (1993). The original concept is a gapless single motif discovery tool based on alternating Markov Chain and Monte Carlo sampling steps (MCMC). It is based on a probabilistic method of finding the best subsequence from all the input sequences which, when aligned, have a maximized consensus alignment score. It has been the

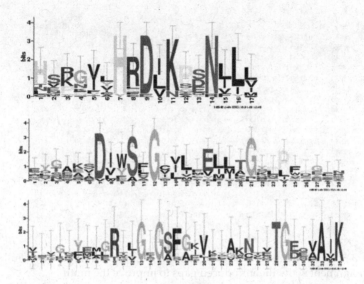

Figure 3-3. MEME motifs found for three sets of plant *Arabidopsis* proteins. The sequence logos show the more conserved, informative residue letters in bold, the height of the letter shows the information content of the alignment in bits.

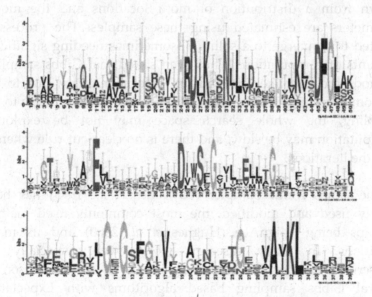

Figure 3-4. GLAM2 motifs found for three sets of plant *Arabidopsis* proteins. The sequence logos show the more conserved, informative residue letters in bold, the height of the letter shows the information content of the alignment in bits.

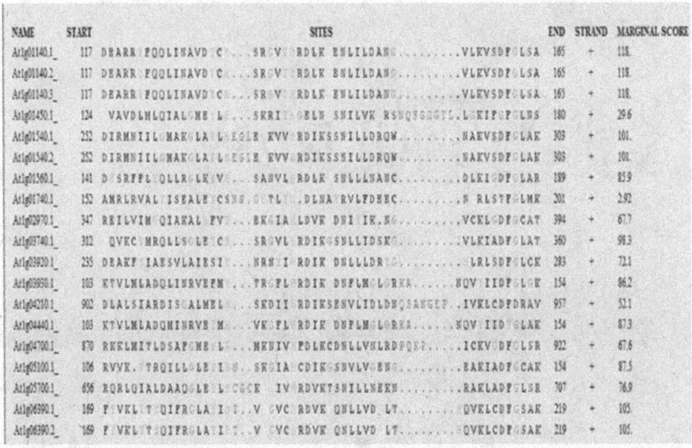

NAME	START	SITES	END	STRAND	MARGINAL SCORE
At1g01140.1_	117	DEARR FQQLINAVD C...SRGV RDLK ENLILDANG.........VLKVSDFGLSA	165	+	118.
At1g01140.2_	117	DEARR FQQLINAVD C...SRGV RDLK ENLILDANG.........VLKVSDFGLSA	165	+	118.
At1g01140.3_	117	DEARR FQQLINAVD C...SRGV RDLK ENLILDANG.........VLKVSDFGLSA	165	+	118.
At1g01450.1_	124	VAVDLMLQIALGME L...SKRI GELN SNILVK RSNQSGSGTL.LGKIFGPGLNS	180	+	29.6
At1g01540.1_	252	DIRMNIILGMAKGLA LEGLE KVV RDIKSSNILLDRQW.........NAKVSDFGLAK	303	+	101.
At1g01540.2_	252	DIRMNIILGMAKGLA LEGLE KVV RDIKSSNILLDRQW.........NAKVSDFGLAK	303	+	101.
At1g01560.1_	141	D SRFFL QLLRGLK V...SANVL RDLK SNLLLNANC.........DLKIGDFGLAR	189	+	85.9
At1g01740.1_	152	ANRLRVAL ISEALE CSNS, TL .DLNA RVLFDENC.........N RLSTFGLMK	201	+	29.2
At1g02970.1_	347	REILVIM QIAKAL FV ...EKGIA LDVK DNI IK.NG.........VCKLGDFGCAT	394	+	67.7
At1g03740.1_	312	QVKC MRQLLSGLE C...SRGVL RDIKGSNLLIDSKG.........VLKIADFGLAT	360	-	98.3
At1g03920.1_	235	DEAKF IAESVLAIESI ...NRN I RDIK DNLLLDR ....LRLSDFGLCK	283	+	72.1
At1g03950.1_	103	KTVLNLADQLINRVEFM ...TRGFL RDIK DNFLMGLGRKA......NQV IIDFGLGK	154	+	86.2
At1g04210.1_	902	DLALSIARDISGALME L...SKDII RDIKSENVLIDLDNQSANGEP..IVKLCDFDRAV	957	-	52.1
At1g04440.1_	103	KTVLMLADQMINRVE M...VKGFL RDIK DNFLMGLGRKA......NQV IID GLAK	154	+	87.3
At1g04700.1_	870	RKKLMITLDSAFGME L...MKNIV PDLKCDNLLVNLRDPQRP.....ICKVGDFGLSR	922	+	67.6
At1g05100.1_	106	RVVK. TRQILLGLE I N..SKGIA CDIKGSNVLVGENG.........EAKIADFGCAK	154	+	87.5
At1g05700.1_	656	RQRLQIALDAAQGLE L CGCK IV RDVKTSNILLNEKN.........RAKLADFGLSR	707	+	76.9
At1g06390.1_	169	F VKL T QIFRGLA I T..V GVC RDVK QNLLVD LT.........QVKLCDFGSAK	219	+	105.
At1g06390.2_	169	F VKL T QIFRGLA I T..V GVC RDVK QNLLVD LT.........QVKLCDFGSAK	219	+	105.

Figure 3-5. GLAM2 multiple alignment of the first motif in Figure 3-4. Full stops in the alignment show the introduced gaps to improve the motif.

basis for a large variety of algorithms. In Gibbs sampling the whole search space is not explored, but random samples are drawn from a distribution of motif locations and the model parameters are estimated using these samples. The process is iterated to converge to a solution, sometimes needing significant amounts of computation. It has been stated that Gibbs sampling methods are less likely to end up in local maxima, due to the random sampling step (MacIsaac & Fraenkel 2006), but due to the sampling, the whole search space may not be explored. Computation may be slow, and there is no clear-cut rule when to stop the iterations.

The Gibbs Motif Sampler (Lawrence *et al.* 1993) has been widely used and modified, the most commonly used method perhaps being AlignAce (Hughes *et al.* 2000) and its many variations. In many cases it works well, especially for transcription factor binding site discovery. A comparison of several Gibbs sampling based algorithms with Expectation Maximum based (and other motif finding methods) is given in

Tompa *et al.* (2005). Another extensive survey of the multitude of published motif finding methods can be found in Sandve & Drabløs (2006).

The overall general conclusion of such comparisons and reviews is that many of the tools complement each other, with no clear winner for analyzing all types of datasets. This clearly argues for the use of multiple motif finders and for the usefulness of utilizing several algorithms in tandem or as an ensemble of classifiers, as discussed further in the next chapter.

3.1.2 *Multiple motif models: LOGOS and MotifRegressor*

Searching for multiple combinations of motifs at varying distances from each other is a much harder combinatorial search problem than finding a single conserved motif common to a set of sequences. The main problems are to limit the search space to get the answer in a reasonable time and to devise the suitable thresholds of accepting or rejecting the co-occurring patterns.

LOGOS (Xing *et al.* 2004) is a Bayesian motif finder, which can search for several simultaneous motifs. A useful observation is that the search space can be limited by the fact that in biological motifs there are often several conserved residues adjacent each other, flanked by a region of adjacent heterogeneous non-conserved residues. Thus single scattered "conserved" residues are more likely to be due to a random alignment in the training set.

The LOGOS motif modelling system formalizes the motif representation into two distinct components: First, the local motif detection and alignment model, which includes the Position Weight Matrices (PMWs) and aligns the sequences to motif candidates using a Hidden Markov Dirichlet-Multinomial (HMDM) model, which can include prior biological knowledge from known motifs. Second, the global Hidden Markov Model (HMM) for simultaneous multiple motif detection is used, rather than models of the background sequence properties and the frequencies and locations of the aligned candidate motifs. For statistics, a Variational Expectation Maximum (VEM) is used. The LOGOS method is stated to outperform MEME and AlignAce in terms of false positives and false negatives. However, this complex modelling system has not become widely used, and MEME appears to remain more popular, perhaps due to its ease of use and sophisticated web services for on-line analysis.

The large number of available *de novo* motif finders makes is difficult to choose the best method for the data at hand. Recently, a flexible motif finding tool comparison platform called MTAP for motif tool selection was introduced by Quest *et al.* (2008). Extensive comparisons, including MEME, AlignAce, GLAM, Weeder *etc.*, indicated that MEME, AlignAce and Weeder work well for transcription factor binding site detection when using RegulonDB (a large database for known gene regulatory regions for the bacterium *Escherichia coli*) for validation. Also, MEME appears to have less false positives than other methods, which may be due to the intrinsic advantage of expectation maximum statistics over the randomized Gibbs sampling based methods.

A useful integrative approach to improve motif finding is to utilize several different motif algorithm in either simultaneously or in tandem (that is, kind of sequential ensemble methods) and to combine the results. As an example, MotifRegressor (Conlon *et al.* 2003) detects motifs using additional information from microarray experiments on gene expression, and is an example of integrated motif discovery. The *de novo* search for motifs is made by MDScan and the obtained large list of motif candidates is then screened by the regression to microarray expression data. Not surprisingly, the method outperforms MEME, because more information is used. This example highlights the benefit of integrated methods using data from different domains, which is becoming a standard approach in integrative bioinformatics.

Often one algorithm finds a short motif in one location, and another algorithm another nearby motif or a longer or overlapping motif. These can usefully be combined to a longer motif or a group of two linked short motifs. This approach was implemented as a web-based service WEBMOTIFS by Romer *et al.*

(2007), in which four algorithms (MEME, AlignAce, Weeder and MDScan) can be run simultaneously and their results combined by clustering the overlapping motifs together. In addition, WEBMOTIFS can do Bayesian motif discovery by comparison to known DNA binding motifs by SVM-based THEME algorithm (MacIsaac *et al.* 2006) utilizing experimental protein binding data. In general, utilizing additional data from laboratory experiments is likely to improve results, as recently shown by W-AlignAce implementation, which combines the Gibbs sampler with gene expression or protein binding experimental data to select the most significant motifs having similar experimental expression or binding profiles (Chen *et al.* 2008).

Other such motif algorithm integration platforms are also starting to appear, *e.g.* MotifToolManager (Phan & Furlotte 2008), which incorporates MEME, AlignAce, MDScan, BioProspector and DME. The most sophisticated one so far appears to be MotifVoter (Wijaya *et al.* 2008), which uses an impressive ensemble of 10 different algorithms, consisting of seven Position Weight Matrix based methods (AlignACE, ANN-Spec, BioProspector, Improbizer, MDScan, MEME, MotifSampler) and 3 other mutated motif based methods (MITRA, Weeder and SPACE). The innovation in the MotifVoter is that the conserved positions in all the overlapping motifs found by different algorithms are used to define the final consensus motif, thus using a maximum amount of information.

The mutated motif methods (also called (l,d) methods) are based on the motif finding problem formulation of Pevzner & Sze (2000), in which a known consensus motif of length l nucleotides is mutated at d positions, *e.g.* a 15 bases long conserved motif with 3 mutations. The task is to find out if a motif finding algorithm can detect such a mutated motif *de novo* in the data set. This is a simplified representation of the known variation in biological DNA motifs, but serves as a useful information-theoretic benchmark for the performance capability of different motif finders.

In overall conclusion of this subchapter, it can be stated that the *de novo* motif finding methods based on sequence data still have a lot of room for improvement. The most likely reason for the inadequate algorithm performance is the lack of relevant information, that is, the local sequence data itself does not completely determine the biological functions of the DNA binding proteins, folding of DNA and RNA to specific structures *etc.*

There are many chromosomal proteins bound to the genomic DNA affecting its function. Also the methylation of specific nucleotides is also largely unknown in the sequences being analysed for motifs, although it is well known that methylation affects DNA transcription to messenger RNA, *e.g.* via promoter activity control. Therefore further significant progress in motif finding is expected to occur after such data can be integrated with the mere text-based data used by most of the current motif finders. The task of the future integrative bioinformatics algorithms is then to fuse such data effectively from different domains of experimental biology. This kind new data fusion based methods are already starting to appear, as discussed above. Only time will tell which of the approaches work best, and how many different algorithms and how many different types of data are needed to be combined to adequately to find the functional motifs in genes and proteins which determine the complexities of gene function and the evolution of genes and proteins.

3.1.3 *Informative k-mers approach*

The k-mers approach is based on words, or short strings selected to be representative of the motif in a collection of sequences. The expected frequency of the k-mer is used to assess its significance. The k-mer based algorithms are often compute-intensive, involving exhaustive listing and counting of words of different sizes and their frequencies in the data set. They are thought to be best suited to simple and uniform ungapped motifs which are highly conserved.

The k-mers are sometimes called n-grams in earlier literature on automatic classification of texts and languages (Damashek 1995), or suffix arrays and this terminology should be remembered when searching the non-biological literature for this topic. A suffix tree algorithm can be used to enumerate all possible substrings of all sizes from the sequences for full counting of frequencies of all k-mers in any position of the sequence. Recently, fast implementations for the suffix array construction have become available (Zhang & Nong 2008) so that their use for larger datasets becomes feasible. For correct statistics of over- and under-represented k-mers, the choice of the "random" control sequence or expected frequencies is critical. An interesting new method in this area is the generalized k-Truncated Suffix Tree (Schulz *et al.* 2008), which creates a k-deep tree for fast linear time' searches of short motifs in DNA (or protein) sequences.

K-mers can be used as classification features for motif discovery in various contexts. The k-mer approach does not focus on specific pre-determined types of sizes of traditionally used sequence motifs, but instead utilizes the large feature set of all possible enumerated k-mers and their frequencies. Thus this approach gives potentially a richer set of classifying data to start with for the classifier. Therefore both feature selection and weighting of data become relatively more important for the classifier performance.

K-mer under- and over-representation has turned 'out to be a useful indicator of specific domains or functions of larger sequence stretches, for which a reasonable number of counts for each k-mer can be obtained for statistics. Examples include over- and under-represented k-mers in non-coding yeast chromosome segments compared to protein coding segments (Hampson *et al.* 2002), and the method of Noble *et al.* (2005), which used the frequencies of the 2772 k-mers of length 1 to 6 as input to a SVM classifier to successfully detect regulatory regions from the human genome. Not surprisingly, the latter study found that the 2-mer "GC" was the most significant k-mer contributing to good classification, as GC-% has long been known to be more abundant in non-coding genome regions.

K-mers are thus useful for large domain detection in genomes, but smaller motifs can also be detected. For example, plant messenger RNA polyadenylation sites (polyadenylation is a process of adding a polymer of adenine nucleotides AAA in the end of messenger RNA during RNA maturation for its full functionality) have been predicted using SVM and other algorithms. Highly informative k-mer features of length 1–3 can locate the messenger RNA polyadenylation site in *Arabidopsis* plant protein coding sequences (Havukkala & VanderLooy 2007). Figure 3-6 shows the distribution of several informative trimers (k-mers of length 3) in an aligned set of sequences around the known polyadenylation site.

The k-mers approach is not limited to short sequence or gene analysis for limited length motifs, but can be used for genome scale analyses as well. Such larger-scale applications of k-mers are starting to appear, *e.g.* Pavlović-Lazetić *et al.* (2009) used k-mers (n-grams) to find and classify genomic islands in bacterial genomes. Poddar *et al.* (2007) also analyzed both genomic and proteomic sequences for informative k-mers to classify organisms and protein families and to mine conserved protein patterns and repeats in genomes. A comprehensive analysis of 5-20 bp k-mers in the human genome was made by Liu *et al.* (2008), including missing 15-mers that could be used for non-human sequence (due to *e.g.* microbial infection) discovery in tissue samples. The over-represented k-mers can be correlated to many annotated features along the genome, and could reveal novel non-coding conserved motifs of unknown biological function. More genomic applications of k-mers are discussed in chapter 4.2.1.

The perceived utility of the informative k-mers as classifiers has led to the development of sequence based kernels for SVM-based classification. In recent work, Mersch *et al.* (2008) used k-mers of length 3 to 6 with SVM to reliably classify protein exon splicing sites. The oligomer feature selection is obtained by an oligokernel, by which the most informative k-mers can be easily

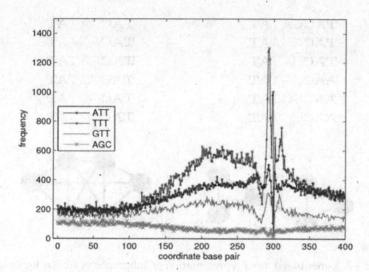

Figure 3-6. The frequency of several k-mers of length 3 around the polyadenylation site of aligned *Arabidopsis* plant protein coding sequences. The polyadenylation site has been aligned at base pair position 300. Several k-mers indicate location and length of different domains around the polyadenylation site. Modified from Havukkala & Vanderlooy (2007). Reproduced by author's personal reuse right requiring no license from Mary Ann Liebert Inc. publication.

obtained. Best classification methods were obtained by a combination of different length k-mers as input to the SVM classification step. Combined k-mer kernels methodology is described in Meinicke *et al.* (2004).

Suffix trees in general hold great promise for string-based enumerative motif finding and many modifications of the suffix tree methods appear possible. One recent example is Weese & Schulz (2007), in which portions of the suffix tree for the k-mers is constructed only when needed using a lazy suffix tree structure. This speeds up the computation and demands less memory as well.

Another way in which k-mers are useful is the recent method called MotifCut (Fratkin *et al.* 2006). It detects motifs by using sets of k-mers to construct a graph in which the vertices are the k-mers in the candidate motifs and the edges are the pairwise similarities between the k-mers. Motifs are found by finding the maximum

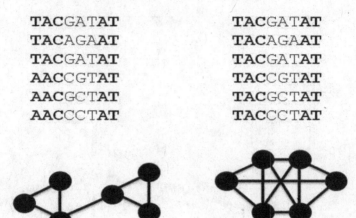

TACGATAT TACGATAT
TACAGAAT TACAGAAT
TACGATAT TACGATAT
AACCGTAT TACCGTAT
AACGCTAT TACGCTAT
AACCCTAT TACCCTAT

Figure 3-7. K-mer based motif representation as subgraphs of similar k-mers. The motif alignment on the left has position dependencies that correspond to two separated subgraphs of k-mer similarities, while the motif alignment on the right corresponds to a compact single graph. The connectors between the k-mers can be based on a selected threshold measuring the distance between the k-mers (*e.g.* 1 or 2 single base mutations necessary for a complete match).

density subgraphs in the graph of all k-mers. This does neatly away with the assumption of independence of each position in aligned motifs (assumed by EM methods, as noted in Figure 3-1 in the beginning of this chapter). Figure 3-7 shows two examples of aligned motifs, one with inter-positional dependencies of nucleotides (corresponding to two separated subgraphs of k-mer similarities), and the other a compact motif without such positional dependencies (corresponding to a single subgraph, suitable for modelling by Position Weight Matrix based motif finders).

This new graph-theoretic approach based on k-mers deserves more attention, especially, because benchmark tests indicated better performance than the other well established methods of MEME, AlignAce and BioProspector. This system should be tested more extensively on more varied datasets to evaluate its general utility. In a similar vein, Kuang *et al.* (2005) used k-mers and protein similarities to build similarity graphs for ranking motifs. The general approach of changing the motif matching from text based alignment to a graph comparison domain is interesting, as graph-based computing algorithms can be very efficient computationally and may explore the feature space in a novel way to find subtle differences between motifs or achieve more reliably the global maximum.

In summary, the informative k-mers is a simple and powerful approach giving a large feature space for the classifiers to explore. This undervalued approach is quickly gaining further popularity due to the recent good results with SVMs using k-mer data as feature input vectors. Extension to methods of using two or more k-mers at specified spacings or gapped k-mers may improve classifier performance even further, provided the search space can be kept small enough to avoid excessive computation times.

The success of this approach is likely due to the biological machinery in cells actually recognizing the small molecular patterns due to specific k-mers, and because the specific sequences of DNA or RNA nucleotides are endowed with specific stiffness, folding, methylation, opening and closing of the double-helix and other important physical properties of the biologically functional DNA. The role of in silico computation of such physical parameters along the genome in functional domain discovery is discussed further in later subchapter 4.3.1. The use of k-mers for global genome analysis is covered further in the subchapter 4.2.1.

3.2 Protein motif discovery

As stated before, protein sequences contain more information per residue than DNA, so that even short informative motifs can be more easily recognized. Also, the proteins are evolutionarily more stable, so that sequences from even remotely related species can be useful in the motif alignments to find the conserved motif. The number of published protein motif finding algorithms is also very large, and a full review is not attempted here. Many of the DNA motif detection algorithms work with modifications for proteins as well, for example MEME covered in the previous subchapter, with a protein motif example. Useful reviews of protein motif algorithm types are those of Yang *et al.* (2005, 2008b) and Ciriello & Guera (2008).

A critical issue for all motif discovery algorithms in general is the ranking and significance measures for the candidate motifs. Motif evaluation metrics are well reviewed in Ferreira & Azevedo (2007). Their survey included 14 different motif significance measures, and the main conclusion was that several significance measures should be used in combination. For strongly conserved motifs most measures work well, but for weakly conserved motifs Z-score or Log-Odd metrics are best. In the case of differentiating closely related motifs with very similar features, the support-based metrics (support metric is based on the number of different sequences where the motif occurs) are most appropriate. Similar recommendations may apply to DNA motif metrics, but a similar motif significance metric survey on DNA motif finders would be useful to standardize the reporting of motif algorithm performance and to aid comparisons between algorithms.

As an example of pattern based motif finders, PROSITE motifs and InterProscan are introduced, together with the various motif databases utilized by InterProScan.

3.2.1 *InterProScan and other traditional methods*

PROSITE motifs were introduced in the earlier subchapter 2.2.1.2. EBI, who hosts the PROSITE database, has also developed the software called InterProScan which can quickly search a database for known PROSITE motif signatures of the InterPro member databases (PROSITE, PRINTS, Pfam, ProDom, SMART, TIGRFAMMs). The PROSITE motifs are best for finding short conserved motifs. The PRINTS motif signatures excel in determining specific closely related sub-family similarities between proteins. The other profiles and also TIGRFAMMs are suitable for identifying more distantly related proteins in divergent super-families. Ideally, for comprehensive *ab initio* analysis, all motif databases should be scanned with InterProScan. The scanning results of InterProScan include in addition to the visualization of the motif locations also tabulations with statistical measures on the significance of matches. The databases scanned by InterProscan are shown in the flow diagram of Figure 3-8 and sample output is shown of an InterProScan run on the webserver in Figure 3-9.

A recently noted methodological problem for protein motif finders is that none of the currently existing multiple alignment algorithms may not actually be suitable for obtaining the "correct" alignments as a starting point for motif evaluation, because these alignment methods have been designed to suit small globular proteins with a compact local homology domain and are not

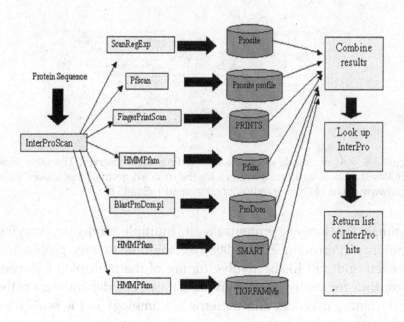

Figure 3-8. The databases scanned by InterProScan. Reproduced from http://www.ebi.ac.uk/2can/tutorials/function/InterProScan2.html.

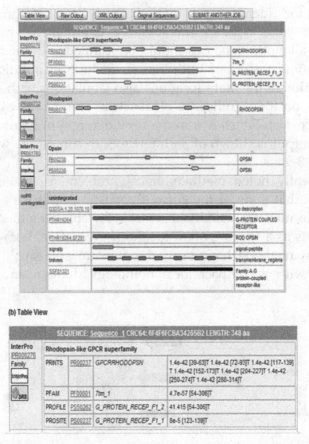

(b) Table View

Figure 3-9. Part of sample output from InterProScan webserver. The table view includes statistical significance values for the detected motifs. Reproduced from http://www.ebi.ac.uk/2can/tutorials/function/InterProScan5.html.

optimal for longer proteins with multiple areas of varying similarity (Perrodou *et al.* 2008). A solution to this problem is awaited and will likely involve tuning of the multiple alignment algorithm for each gene family or dataset, *i.e.* depending on the evolutionary diversity and patterns of homology in the sequences within the dataset.

Another technical issue is that care needs to be taken to choose a suitable Position Weight Matrix (PWM) for determining the existence of the motif that is significantly different from the "random" background sequence. A new scoring method based on both the information content of the motif and its statistical expectation in relation to the background sequences for the match between the motif and the PWM has been proposed by Pan & Phan (2008).This method has not yet been tested extensively to see whether it is truly superior than the traditional simpler approach.

A third technical issue is the problem of deciding what kind of PWM should one construct for the detected motif to screen new sequences with. Ideally, experimental data should be used to validate the PWM used to select motifs from unknown sequences. One recent suggestion how to achieve this is to optimize the PWM using a genetic algorithm based on iterative feedback from experimental microarray or binding data (Li *et al.* 2007). This kind of approach is appealing, but obviously can be used only when some additional external data is available for the iterative optimization of the PWM. Ideally, one should try to collect additional information about the set of sequences anyway, *e.g.* by making a phylogenetic analysis to obtain further related sequences for pooled analysis or by retrieving other possible annotations available for the sequences.

3.2.2 *Protein k-mer and other string based methods*

Amino acid k-mers can be considered informative features just as well as DNA k-mers explored in the previous subchapter 3.1.1, because short amino acid domains affect hydrophobicity, protein fold formation, bulkiness of the amino acid chain and other crucially important physico-chemical factors recognized by the metabolic and regulatory systems of living cells. K-mers therefore should provide relevant and distinguishing features for short domains in *de novo* protein motif discovery as well as in protein classification, provided enough data is available for evaluation the significance of k-mer frequency deviations.

Many of the k-mer DNA motif finding approaches could be adapted to protein motif analysis as well, but this has not .been explored much yet. One reason may be that sufficient amounts of data may not be always available. However, with new large proteomic datasets becoming available, protein k-mers also deserve more attention.

An early paper of using k-mers (also called n-grams) for protein classification is Cheng *et al.* (2004), starting from classical text classification methods. More recently, protein k-mers of length 4 were used to classify SCOP protein families and GPCR receptor protein subfamilies (Yang *et al.* 2008a). Informative k-mers were ranked by frequency or an entropy measure and then the feature vector of the frequencies of the selected k-mers used in an SVM classifier. Significantly, the method seems to outperform the traditional BLAST similarity search by a clear margin, and also betters the spectrum kernel approach.

Spectrum kernels (Leslie *et al.* 2002) can be set up to utilize the complete set of computable k-mers, and although they perform well in general, there may be too much noisy information in the complete set of k-mers that hampers classification performance, at least in the comparison test made by Cheng *et al.* (2004). A full survey of various kernel methods is not attempted here, but a good review is available (Schölkopf *et al.* 2004) and kernel methods related to SVMs are rapidly gaining popularity in bioinformatics. Filtering intelligently the set of k-mers to be used as discriminatory features appears essential for further improvement in k-mer based motif finders and classifiers.

An interesting full proteome peptide word vocabulary analysis based on short 6–12 mer patterns in amino acid sequences was made by Gatherer (2007). Using word enumeration methods from text processing algorithms, protein families and related proteins in other proteomes could be discovered. In a related manner, Wang *et al.* (2005) identified transcription factor binding motifs by enumerating distinguishing k-mers and learning a grammar.

Their algorithm WordSpy was then used successfully to discover core promoters for non-coding microRNAs (Zhou *et al.* 2007). It is clear that linguistic methods, suitably modified, can contribute new aspects to the use of k-mers in motif discovery algorithms.

3.3 Genetic algorithms, particle swarms and ant colonies

Genetic algorithms (GA), particle swarm optimization (PSO) and Ant Colony Optimization (ACO) are popular in artificial intelligence and machine learning applications in the engineering and visual information processing fields. Recently they have been also applied to DNA and protein motif detection as well, and are becoming increasingly popular in other areas of bioinformatics. A full survey of this burgeoning field is not possible here, so that only exemplary bioinformatics related pattern finding algorithms are presented here.

3.3.1 *Genetic algorithms*

Genetic algorithms (GA) have a long history and a huge literature in a large variety of technical disciplines. The method is based on assembling variable sets of features to long linear feature vectors called chromosomes, and crossing two chromosomes at each iteration (generation) to produce variation in the feature vectors in a process akin to natural evolution and crossing over of gene variants between paternal and maternal genetic material. In spite of the high computational cost, they are becoming more widely adopted due to the increasing power of modern desktop and laptop computers. In general GA methods converge well, but the global maximum is not guaranteed. A recent useful review of Genetic Algorithms for bioinformatics is in Pal *et al.* (2006).

Already in 1996 (Koza & Andre 1996) GA was used to discover protein motifs, but the method was not widely adopted, due to many more competitive traditional algorithms being available. More recent work includes Wei & Jensen (2006) who developed a genetic algorithm to detect cis-regulatory elements. Chan *et al.* (2008) reported the genetic algorithm GALF-P for transcription factor binding site discovery by improved representation of the features in the artificial chromosomes combined with Local Filtering. GALF-P was stated to outperform MEME and BioProspector, and may be considered the current state-of-art GA system for transcription factor binding site detection.

For the more difficult problem of composite binding site motifs (*i.e.* two or more motifs responsible for the activation of a promoter, which is well known to occur in gene regulation), a new GA algorithm of Fogel *et al.* (2008) shows promise. The advanced feature in their algorithm is that the genetic variation parameters are modified during evolution depending on the stage of motif discovery. Another algorithm for the composite motif problem is that of Lones & Tyrrell (2007), which mimics natural selection even more by including co-evolutionary selection pressures in the population screening process.

Another novel direction is to use the genetic algorithm in a multi-objective setting. Kaya (2007) used GA to optimize simultaneously motif length (longer motifs are more desirable) and statistical support for the significance of the motif (longer motifs normally have more variants and less support). Results were encouraging when compared to MEME, AlignAce and Weeder, but the general applicability of this method with different datasets remains to be explored. Rajapakse *et al.* (2007) also used multi-objective evolutionary algorithm (MOEA) for selecting peptide motifs binding to immunologically important MHC class II molecules.

Genetic algorithms appear to be useful for SVM kernel optimization as well, especially for the string kernels relevant to motif finding. Håndstad *et al.* (2007) produced a well-performing SVM motif kernel called GP kernel by evolving the kernel features according to the sequence similarities. It performed well on the SCOP protein family database, but more comparisons with different datasets and with *e.g.* MEME and AlignAce are needed.

The general problem in genetic algorithms seems to be to find an optimal representation and selection of features in the chromosomes as well as to select suitably tuned mutation and crossing-over frequencies. Ideally, the mutation and crossing-over frequencies would be adjusted dynamically, as done in Fogel *et al.* (2008) discussed above. The other issue is the computational cost, which is often a limiting factor for the size of the chromosome populations and the number of selection and crossing-over iterations that can be performed. This basic limitation applies also to the PSO and ACO algorithms, both of which are discussed below.

3.3.2 *Particle swarm optimization*

Particle swarm optimization (PSO) algorithms explore the search space with a large number of semi-independent agents, which coordinate their search with each other, like individuals in a bird flock or a fish school. Introduced by Kennedy & Eberhart (1995), they have become popular in engineering and recently also in bioinformatics. A good survey of PSO methods in general is in Banks *et al.* (2007, 2008). Bioinformatics algorithms utilizing PSO are not as numerous as those for ACO. A recent review of bioinformatics applications of PSO is available in Das *et al.* (2008) and includes also a review of ACO applications.

In particle swarm optimization one of the main advantages is that the search space is more thoroughly searched from a larger area of multiple starting points, so that the global maximum is more likely to be found. Also, the computation is divided into smaller chunks, which can be separately processed in a distributed grid environment with multiple computers.

An early paper for PSO use in protein motif detection was Chang *et al.* (2004), and protein multiple alignments were optimized using a PSO in Hsiao *et al.* (2005). An interesting combination of Genetic Algorithms and PSO was suggested by Miranda & Fonseca (2002), and this has been gathering a following in the electric power industry research, but could be used in bioinformatics data analysis as well. Other new hybrid methods include GA or PSO combined with Greedy Randomized Adaptive Search Procedure GRASP for clustering (Marinakis *et al.* 2008a, 2008b).

Such hybrid methods have been little used in bioinformatics, with the exception of microarray data analysis. One of these interesting hybrid methods is the GA-IBPSO which was used for gene selection from microarray using genetic algorithm together with improved binary particle swarm optimization, with promising results (Yang *et al.* 2008a). A further example is the use of GA and PSO in tandem for gene selection, followed by SVM classification (Li *et al.* 2008). In this case the more efficient gene selection process appears to introduce a better quality feature set to the SVM. Biclustering of microarray has also been done by GA-PSO combination (Xie *et al.* 2007).

Motif analysis can also be performed using a hybrid GA-PSO. A recent paper by Chang *et al.* (2008) detected transcription factor binding sites in promoters by modifying Position Weight Matrix by GA and using PSO to explore the feature space. The method appears to outperform MEME and Gibbs Sampler and seems worth further development. It will be interesting to see, how such hybrid methods will work for more challenging bioinformatics problems and whether the increased computational cost is worth the improvement in performance.

3.3.3 *Ant colony optimization*

Ant colony optimization (ACO) mimics the way ants leave pheromone tracks for other ant individuals to follow. The search agents are designed to leave track information for the other agents in the algorithm to follow. ACO was developed by Dorigo, and used early on for job scheduling (Colorni *et al.* 1994) and industrial process control. There is a voluminous literature on ACO, including many books; a good recent summarizing book by the developer of the algorithm is Dorigo & Stützle (2004). For a review on bioinformatics domain ACO algorithms, see Das *et al.* (2008).

Recent motif finders utilizing ACO include finding immunological MHC II class binding motifs (Karpenko *et al.* 2005). This method needs, however, improvement to stay competitive with the alternative probabilistic methods which are also improving rapidly (Zhang *et al.* 2009).

In addition to short sequence motif finding, ACO has been recently adapted to matching and aligning longer sequences in the multiple alignment problem. Guinand & Pigne (2008) and Gondro & Kinghorn (2007) devised prototype multiple sequence aligners, but it is not clear, if they will be competitive in terms of alignment quality. Computational speed is likely to be inferior to the standard deterministic multiple alignment methods, which do not have random components in their search for matching sequence segments.

The ACO seems well suited for 3-dimensional protein folding problems with local structural motifs, because the search space is large and complicated, and is explored efficiently by the ACO, at least in the implementation of Shmygelska & Hoos (2005). It remains to be seen, if the computational demands can be met for the more complex large protein folding problems.

3.4 Sequence visualization

Visualization of sequences as such is an important and large sub-discipline in bioinformatics, as in any data mining discipline dealing with long linear symbol strings. There are many biology-specific tools that have been developed for visualizing multiple alignments and motif locations in sequences. Such tools are normally integrated in the software packages for various search algorithms, *e.g.* the BLAST hit summaries at NCBI and phylogeny trees in the PhyLip package. Standalone tools for general use include T-coffee for flexible multiple alignments and sequence logos for summarizing motif information. They are valuable research tools for visual browsing of results and *ad hoc* interactive analysis.

Sequence logos were developed by Schneider (1990), and have become very popular and are now also available as a web service WebLogos (Crooks *et al.* 2004). Sequence logos are an established visualization method for aligned short DNA and protein sequences. We saw an example of sequence logos in the beginning of this chapter and in subchapter 3.1.1. More advanced features have been added to the basic visualization, the most recent being the Phylo-mLogo, which can display alignments of up to thousands of sequences interactively (Shih *et al.* 2007).

An interesting and useful extension to WebLogo is the two-sample logo (Vacic *et al.* 2004), which can show the salient differences between two aligned motifs very nicely. An example is shown in Figure 3-10. Furthermore, the information extracted by two-sample logo has been used successfully as summarized features for input to SVM classifier for splice site recognition (Xu *et al.* 2007). This method of summarizing two-sample logo data for classifiers would apply in principle for any two-motif comparisons. The increased performance found by Xu *et al.* appears to derive from the pre-processing step, which in essence filters the data to amplify the essential differences between the two motifs. It also indicates the biological relevance of the individual nucleotides in the motif for splicing the DNA. This general approach warrants further study for any hard to distinguish motif classification, as it is always worth to first find a data presentation that shows the putative differences clearly, before selecting the classificatory features and tuning the classification algorithm itself.

T-COFFEE is a popular multiple alignment tool (Notredame *et al.* 2000), further extended to M-COFFEE (Wallace *et al.* 2006) to assemble several multiple alignments from different sources together. There is also an accurate non-coding RNA multiple aligner called R-COFFEE (Morettti *et al.* 2008). It is one of the best RNA aligners and takes into account both conserved hairpin structures (using the Vienna Package RNAplfold) and sequence consensi in making the final consensus sequence. R-COFFEE is

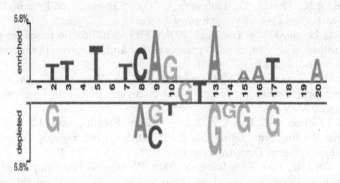

Figure 3-10. Comparison of two sets of aligned sequences by the Two-Sample Logo. The top logo letter heights indicate the level of enrichment in the nucleotides in the alternatively spliced sequences compared to the regularly spliced sequences. The bottom graph shows the same for the depleted nucleotides. The motif is centered around the GT splicing site. From Vacic *et al.* (2006). Reproduced by permission from Oxford University Press.

also available as a web service at the T-coffee website www.tcoffee.org. It exemplifies the trend to data analysis and visualization methods that combine various types of data (*e.g.* sequence data and structural data) to better classify and visualize the salient features of the dataset. Its implementation shows that for integration of the various data types and feature analyzers, the modularity and compatibility of the different algorithms is essential, so that integrated methods can be easily put together and modified and extended as necessary.

References

Bailey, T.L., Williams, N., Misleh, C. & Li, W.W. 2006. MEME: discovering and analyzing DNA and protein sequence motifs. *Nucleic Acids Research 34:* W369–W373.

Banks, A., Vincent, J. & Anyakoha, C. 2007. A review of particle swarm optimization. Part I: background and development. *Natural Computing 6(4):* 467–484.

Banks, A., Vincent, J. & Anyakoha, C. 2008. A review of particle swarm optimization. Part II: hybridisation, combinatorial, multicriteria and constrained optimization, and indicative applications. *Natural Computing 7(1):* 109–124.

Björklund, Å.K., Ekman, D., Elofsson, A. 2006. Expansion of Protein Domain Repeats. *PLoS Computational Biology 2(8): e114*

Chan, T.M., Leung, K.S. & Lee, K.H. 2008. TFBS identification based on genetic algorithm with combined representations and adaptive post-processing. *Bioinformatics 22(13): 1577–1584.*

Chang, B.C.H, Ratnaweera, A., Halgamuge, S.A. & Watson, H.C. 2004. Particle swarm optimization for protein motif discovery. In: *Genetic Programming for Evolvable Machines 5: 203–214.*

Chang, X-Y., Zhou, C-G. Liu, Y-W. & Hu, P. 2006. Identification of Transcription Factor Binding Sites Using GA and PSO. *Sixth International Conference on Intelligent Systems Design and Applications (ISDA). 1: 473–480.*

Chen, X., Guo, L., Fan, Z. & Jiang, T. 2008. W-AlignACE: an improved Gibbs sampling algorithm based on more accurate position weight matrices learned from sequence and gene expression/ChIP-chip data. *Bioinformatics 24(9): 1121–1128.*

Cheng, B.Y.M., Carbonell, J.G. & Klein-Seetharaman, J. 2004. Protein Classification Based on Text Document Classification Techniques. *Proteins: Structure, Function and Bioinformatics 58: 955–970.*

Ciriello, G. & Guerra, C. 2008. A review on models and algorithms for motif discovery in protein-protein interaction networks. *Briefings in Functional Genomics and Proteomics 7(2): 147–156.*

Colorni, A., M. Dorigo, V. Maniezzo and M. Trubian. 1994. Ant System for Job Scheduling. *Belgian Journal of Operations Research, Statistics and Computer Science 34: 36–53.*

Conlon, E.M., Liu, X.S., Lieb, J.D. & Liu, J.S. 2003. Integrating regulatory motif discovery and genome-wide expression analysis. *Proc Natl Acad Sci U S A. 100(6): 3339–3344.*

Crooks, G.E., Hon, G., Chandonia, J.M. & Brenner, S.E. 2004. WebLogo: A sequence logo generator. *Genome Research 14: 1188–1190.*

Damashek M. 1995. Gauging Similarity with n-Grams: Language-Independent Categorization of Text. Science.267(5199): 843–848.

Das, M.K. & Dai, H.-K. 2007. A survey of DNA motif finding algorithms. *BMC Bioinformatics 8(Suppl 7): S21.*

Das, S., Abraham, A. & Konar, A. 2008. Swarm Intelligence Algorithms in Bioinformatics. *Chapter 4, pp. 113–147 in: Computational Intelligence in Bioinformatics Vol. 94, Springer.*

Do, C.B. & Batzoglou, S. 2008. What is the expectation maximization algorithm? *Nature Biotechnology 26: 897 – 899.*

Dorigo, M. & Stützle, T. 2004. Ant Colony Optimization. MIT Press, 319 pp.

Ferreira, P.G. & Azevedo, P.J. 2007. Evaluating deterministic motif significance measures in protein databases. *Algorithms Mol Biol. 2: 16.*

Fogel, G.B., Porto, V.W., Varga, G., Dow, E.R., Craven, A.M., Powers, D.M., Harlow, H.B., Su, E.W., Onyia, J.E. & Su, C. 2008. Evolutionary computation for discovery of composite transcription factor binding sites. *Nucleic Acids Res. 36(21): e142.*

Fratkin, E., Naughton, B.T., Brutlag, D.L. & Batzoglou, S. 2006. MotifCut: regulatory motifs finding with maximum density subgraphs. *Bioinformatics* 15: e150–157.

Frith, M.C., Saunders, N.F.W., Kobe, B. & Bailey, T.L (2008) Discovering Sequence Motifs with Arbitrary Insertions and Deletions. PLoS Computational Biology 4(5): e1000071.

Gatherer, D. 2007. Peptide Vocabulary Analysis Reveals Ultra-Conservation and Homonymity in Protein Sequences. *Bioinformatics and Biology Insights 1: 101–126.*

Gondro, C. & Kinghorn, B. P. 2007. A simple genetic algorithm for multiple sequence alignment. *Genet. Mol. Res. 6(4): 964–982.*

Guinand, F. & Pigné, Y. 2008. An Ant-based Model for Multiple Sequence Alignment. *In: Lirkov, I. Margenov, S. & Wasniewski, J. (eds.) Proceedings of the 6th SIAM International Conference on "Large-Scale Scientific Computations" (LSSC'07). Springer-Verlag, LNCS 4818: 553–560.*

Habib, N., Kaplan, T., Margalit, H. & Friedman, N. 2008. A Novel Bayesian DNA Motif Comparison Method for Clustering and Retrieval. *PLoS Computational Biology 4(2): e1000010 doi:10.1371/journal.pcbi.1000010.*

Hampson, S., Kibler, D. & Baldi, P. 2002. Distribution patterns of over-represented k-mers in non-coding yeast DNA. *Bioinformatics 18(4): 513–528.*

Håndstad T, Hestnes AJ, Saetrom P. 2007. Motif kernel generated by genetic programming improves remote homology and fold detection. BMC *Bioinformatics 8: 23.*

Havukkala, I. &Vanderlooy, S. 2007. On the Reliable Identification of Plant Sequences Containing a Polyadenylation Site. *Journal of Computational Biology 14(9): 1229–1245.*

Hsiao, Y.T., Chuang, C.L., Jiang, J.A. 2005. Particle swarm optimization approach for multiple biosequence alignment. *In: Proceedings of the IEEE international workshop on genomic signal processing and statistics 2005, Rhode Island, USA, 22–24 May 2005.*

Hughes, J.D., Estep, P.W., Tavazoie, S. & Church, G.M. 2000. Computational identification of cis-regulatory elements associated with groups of functionally related genes in Saccharomyces cerevisiae. *Journal of Molecular Biology 296(5):1205–1214.*

Karpenko, O., Shi, J. & Dai, Y. 2005. Prediction of MHC class II binders using the ant colony search strategy. *Artificial Intelligence in Medicine 35(1): 147–156.*

Kaya, M. 2007. Motif Discovery Using Multi-Objective Genetic Algorithm in Biosequences. *pp. 320–331. In: Advances in Intelligent Data Analysis VII, Springer.*

Kennedy, J. & Eberhart, R.C. 1995. Particle Swarm Optimization", *Proceedings, IEEE International Conference on Neural Networks, Perth, Australia.*

Klepper, K., Sandve, G.K., Abul, O., Johansen, J. & Drablos, F. 2008. Assessment of composite motif discovery methods. *BMC Bioinformatics 9: 123.*

Koza, J.R. & Andre, D. 1996. Automatic Discovery of Protein Motifs Using Genetic Programming. *In Evolutionary Computation: Theory and Applications. Edited by: Yao X. World Scientific.*

Kuang, R., Weston, J., Noble, W.S., Leslie, C. 2005. Motif-based protein ranking by network propagation. *Bioinformatics 21(19): 3711–3718.*

Lawrence, C.E., Altschul, S.F., Boguski, M.S., Liu, J.S., Neuwald, A.F., et al. 1993.Detecting subtle sequence signals: a Gibbs sampling strategy for multiple alignment. *Science 262: 208–214.*

Leslie, C., Eskin, E. & Noble, W.S. 2002. The spectrum kernel: A string kernel for SVM protein classification. *Proceedings of the Pacific Symposium on Biocomputing 7: 566–575.*

Li L, Liang Y, Bass RL. 2007. GAPWM: a genetic algorithm method for optimizing a position weight matrix. *Bioinformatics 23(10): 1188–1194.*

Li, S., Wu, X. & Tan, M. 2008. Gene selection using hybrid particle swarm optimization and genetic algorithm. *Soft Computing 12(11): 1039–1048;*

Liu, Z., Venkatesh, S.S. & Maley, C.C. 2008. Sequence space coverage, entropy of genomes and the potential to detect non-human DNA in human samples. *BMC Genomics. 9: 509.*

Lones, M.A. & Tyrrell, A.M. 2007. A co-evolutionary framework for regulatory motif discovery. *IEEE Congress on Evolutionary Computation pp. 3894–3901.*

MacIsaac, K.D. & Fraenkel, E. 2006. Practical strategies for discovering regulatory DNA sequence motifs. *PLoS Comput Biol. 2(4): e36.*

MacIsaac, K.D., Gordon, D.B., Nekludova, L., Odom, D.T., Schreiber, J., Gifford, D.K., Young, R.A. & Fraenkel, E. 2006. A hypothesis-based approach for identifying the binding specificity of regulatory proteins from chromatin immunoprecipitation data. *Bioinformatics 22(4): 423–429.*

Marinakis, Y., Marinaki, M., Doumpos, M., Matsatsinis, N.F. & Zopounidis, C. 2008a. A hybrid stochastic genetic-GRASP algorithm for clustering analysis. *Operational Research 8(1): 33–46.*

Marinakis, Y., Marinaki, M., Doumpos, M. & Matsatsinis, N.F. 2008b. A Hybrid Clustering Algorithm Based on Multi-swarm Constriction PSO and GRASP. *DaWaK 2008: 186–195.*

Meinicke, P., Tech, M., Morgenstern, B. & Merkl, R. 2004. Oligo kernels for datamining on biological sequences: a case study on prokaryotic translation initiation sites. *BMC Bioinformatics 5: 169.*

Mersch, B., Gepperth, A., Suhai, S. & Hotz-Wagenblatt, A. 2008. Automatic detection of exonic splicing enhancers (ESEs) using SVMs. *BMC Bioinformatics 9: 369.*

Miranda, V. & Fonseca, N. 2002. EPSO - best-of-two-worlds meta-heuristic applied to power system problems. *Proceedings of the 2002 Congress on, Evolutionary Computation, pp. 1080–1085.*

Moore, A.D., Björklund, A.K., Ekman, D., Bornberg-Bauer, E., Elofsson, A. 2008. Arrangements in the modular evolution of proteins. Trends in Biochemical Sciences 33(9): 444–451.

Moretti, S., Wilm, A., Higgins, D.G., Xenarios, I. & Notredame, C. 2008. R-Coffee: a web server for accurately aligning noncoding RNA sequences. Nucleic Acids Research 36: W10–133.

Noble, W.S., Kuehn, S., Thurman, R., Yu, M. & Stamatoyannopoulos, J. 2005. Predicting the in vivo signature of human gene regulatory sequences. *Bioinformatics 21(Suppl 1): i338–43.*

Notredame, Higgins & Heringa, 2000. T-Coffee: A novel method for multiple sequence alignments. *Journal of Molecular Biology 302: 205–217.*

Pal, S.K., Bandyopadhyay, S. & Ray, S.S. 2006. Evolutionary computation in bioinformatics: a review. *IEEE Transactions on Systems, Man, and Cybernetics, Part C: Applications and Reviews, 36(5): 601–615.*

Pan, Y. & Phan, S. 2008. Positional weight matrix as a sequence motif detector. Chapter 14, pp. 421–440. In: *Oligonucleotide Array Sequence Analysis, Editors: M.K. Moretti and L.J. Rizzo, Nova Science Publishers, Inc.*

Pavlović-Lazetić, G.M., Mitić, N.S. & Beljanski, M.V. 2009. n-Gram characterization of genomic islands in bacterial genomes. *Comput Methods Programs Biomed. 93(3): 241–256.*

Perrodou, E., Chica, C., Poch, O., Gibson, T.J. & Thompson, J.D. 2008. A new protein linear motif benchmark for multiple sequence alignment software. *BMC Bioinformatics.9: 213.*

Pevzner, P.A. & Sze, S.H. 2000. Combinatorial approaches to finding subtle signals in DNA sequences. *Proc. Int. Conf. Intell. Syst. Mol. Biol. 8: 269–278.*

Phan, V. & Furlotte, N.A. 2008. Motif Tool Manager: a web-based framework *Bioinformatics 24(24): 2930–2931.*

Poddar, A., Chandra, N., Ganapathiraju, M., Sekar, K., Klein-Seetharaman, J., Reddy, R. & Balakrishnan, N. 2007. Evolutionary insights from suffix array-based genome sequence analysis. *Journal of Bioscience 32(5): 8718–81.*

Quest, D., Dempsey, K., Shafiullah, M., Bastola, D & Ali, H. 2008. MTAP: the motif tool assessment platform. *BMC Bioinformatics 9(Suppl 9): S6.*

Rajapakse M,, Schmidt B, Feng L & Brusic V. 2007. Predicting peptides binding to MHC class II molecules using multi-objective evolutionary algorithms. *BMC Bioinformatics 8: 459.*

Redhead, E. & Bailey, T.L. 2007. Discriminative motif discovery in DNA and protein sequences using the DEME algorithm. BMC Bioinformatics 8: 385.

Romer, K.A., Kayombya, G.R. & Fraenkel, E. 2007. WebMOTIFS: automated discovery, filtering and scoring of DNA sequence motifs using multiple programs and Bayesian approaches. *Nucleic Acids Research 35(Web Server issue): W217–20.*

Sandve, G.K. & Drabløs, F. 2006. A survey of motif discovery methods in an integrated framework. *Biology Direct 1: 11.*

Schmidt, E.E. & Davies, C.J. 2008. The origins of polypeptide domains. *BioEssays 29(3): 262–270.*

Schneider, T.D. & Stephens, R.M. 1990. Sequence Logos: A New Way to Display Consensus Sequences. *Nucleic Acids Res. 18:6097–6100.*

Schölkopf, B., Tsuda, K. & Vert, J.-P. 2004. Kernel Methods in Computational Biology. 410 pp. MIT press.

Schulz, M.H., Bauer, S. & Robinson, P.N. 2008. The generalised k-Truncated Suffix Tree for time-and space-efficient searches in multiple DNA or protein sequences. *Int J Bioinform Res Appl.* 4(1): 81–95.

Shih, A.C., Lee, D.T., Peng, C.L. & Wu, Y.W. 2007. Phylo-mLogo: an interactive and hierarchical multiple-logo visualization tool for alignment of many sequences. *BMC Bioinformatics* 24(8): 63.

Shmygelska, A. & Hoos, H.H. 2005. An ant colony optimisation algorithm for the 2D and 3D hydrophobic polar protein folding problem. *BMC Bioinformatics* 6: 30.

Tompa, M., Li, N., Baileym T.L., Church, G.M., De Moor, B., Eskin, E. et al. 2005. Assessing computational tools for the discovery of transcription factor binding sites. *Nature Biotechnology* 23(1):137–144.

Vacic, V., Iakoucheva, L.M. & Radivojac, P. 2006. Two Sample Logo: a graphical representation of the differences between two sets of sequence alignments. *Bioinformatics* 22(12): 1536–1537.

Wallace, I.M., O'Sullivan, O., Higgins, D.G. & Notredame, C. 2006. M-Coffee: combining multiple sequence alignment methods with T-Coffee. Nucleic Acids Research 34(6): 1692–1699.

Wang, G., Yu, T. & Zhang, W. 2005. WordSpy: identifying transcription factor binding motifs by building a dictionary and learning a grammar. *Nucleic Acids Res.* 33: W412–416.

Weese, D. & Schulz, M.H. 2007. Efficient String Mining under Constraints via the Deferred Frequency Index". In: Perner, P. (ed.) Proceedings of the 8th Industrial Conference on Data Mining (ICDM'08), LNAI 5077, pp. 374–388.

Wei, Z., Jensen, S.T. 2006. GAME: detecting cis-regulatory elements using a genetic algorithm. *Bioinformatics* 24(3): 341–349.

Wei, W. & Xiao-Dan Yu, X.-D. 2007. Comparative Analysis of Regulatory Motif Discovery Tools for Transcription Factor Binding Sites. *Genomics, Proteomics & Bioinformatics* 5(2): 131–142.

Wijaya, E., Yiu, S.M., Son, N.T., Kanagasabai, R. & Sung, W.K. 2008. MotifVoter: a novel ensemble method for fine-grained integration of generic motif finders. *Bioinformatics* 24(20): 2288–2295.

Xie, X., Mikkelsen, T.S., Gnirke, A., Lindblad-Toh, K., Kellis, M. & Lander, E.S. 2007. Systematic discovery of regulatory motifs in conserved regions of the human genome, including thousands of CTCF insulator sites. Proc Natl Acad Sci U S A. 104(17): 7145–50.

Xie, B., Chen, S. & Liu, F. 2007. Biclustering of Gene Expression Data Using PSO-GA Hybrid. *The 1st International Conference on Bioinformatics and Biomedical Engineering, 2007. ICBBE 2007. 6–8 July 2007 pp. 302–305.*

Xing, E.P., Wu, W., Jordan, M.I. & Karp, R.M. 2004. Logos: a modular bayesian model for de novo motif detection. Journal of Bioinformatics and Computational Biology 2(1): 127–154.

Xu, S., Ma, F. & Tao, L. 2007. Learn from the Information Contained in the False Splice Sites as well as in the True Splice Sites using SVM. *In: Proceedings of the International Conference on Intelligent Systems and Knowledge Engineering (ISKE 2007) October 15–16, 2007, Chengdu, China*

Yang, C-S., Chuang, L-Y., Ho, C-H. & Yang, C-H. 2008a. Microarray Data Feature Selection Using Hybrid GA-IBPSO. *Lecture Notes in Electrical Engineering 6: 243–253.*

Yang, J., Deogun, J.S. &, Sun, Z. 2005. A New Scheme for Protein Sequence Motif Extraction. *Proceedings of the 38th Hawaii International Conference on System Sciences – Track 9, p. 280a.*

Yang, W.-Y. & Lu, B.-L. 2008c. String Kernels with Feature Selection for SVM Protein Classification. *In: Series on Advances in Bioinformatics and Computational Biology. Eds. Brazma, A, Miyano, S & Akutsu, T. Vol. 6, pp. 9–18. World Scientific Press.*

Yang, Y., Lu, B.-B. & Yang, W.-Y. 2008b. Classification of Protein Sequences Based on Word Segmentation Methods *In: Series on Advances in Bioinformatics and Computational Biology. Eds. Brazma, A, Miyano, S & Akutsu, T. Vol. 6, pp. 177–186. World Scientific Press.*

Zhang, H., Lundegaard, C. & Nielsen, M. 2009. Pan-specific MHC class I predictors: a benchmark of HLA class I pan-specific prediction methods. *Bioinformatics 25(1): 83–89.*

Zhang, S. & Nong, G. 2008. Fast and Space Efficient Linear Suffix Array Construction. *Data Compression Conference, 2008, IEEE Proceedings, pp. 553–553.*

Zhou, X., Ruan, J., Wang, G. & Zhang, W. 2007. Characterization and Identification of MicroRNA Core Promoters in Four Model Species. *PLoS Comput Biol 3(3): e37.*

Chapter 4

Global Pattern Discovery and Comparing Genomes

While local pattern search and comparison of DNA and protein sequences by Smith–Waterman algorithm and its much faster alternative BLAST are well established standard tools in bioinformatics, there are still many challenges to scale up these methods to global pattern discovery and similarity comparisons of complete genomes and proteomes. Genome and proteome comparisons between different species and individuals are necessary for phylogenetic analyses as well as finding out the differences between individuals of the same species for the purposes of *e.g.* plant and animal breeding, personalized medicine and detection of new dangerous forms of pathogens *etc.*

In general, such methods can be divided into **alignment-based methods** and **alignmentless methods**. Alignment based methods rely on string or k-mer analysis of query and target sequence to look for best matching stretches to build the best possible global alignment of the whole sequence and adding of gaps to the alignments to improve the matching sequences. During evolution some genome segments may have reversed and/or moved to another location in the chromosome, which has to be taken into account. The total percentage of aligned identical nucleotides is an overall measure of genome similarity. In contrast, alignmentless methods match and compare some other informative measures either along the sequence or for the global sequence to derive a global similarity measure between the query and the target.

4.1 Alignment-based methods

4.1.1 *Pairwise genome-wide search algorithms: LAGAN, AVID etc.*

Because standard global alignment programs like Needleman–Wunsch algorithm are too slow to align long genome sequences, faster alternative methods have been necessarily developed. All are based on first finding some anchor points of local similarities at various locations in the genomes and then using selected anchors as starting points to refine and extend the alignments to cover all the areas between the anchoring points. Earlier methods like MUMmer or AVID rely on long exactly matching words as anchors. These methods were followed by the further improved LAGAN software (Brudno *et al.* 2002, 2003) which has proved to be an efficient and widely used method in genomics.

LAGAN finds diagonally located anchors of local alignments based on many short partially matching strings using the CHAOS algorithm (Brudno *et al.* 2002). The intervening stretches are then aligned by traditional methods. The procedure principle is shown in Figure 4-1. In this way the alignment problem is subdivided into more manageable smaller tasks that do not need so much computer memory and which can even be parallelized on separate CPUs or networked computers, if necessary.

4.1.2 *Multiple alignment methods: MLAGAN, MAVID, MULTIZ etc.*

MLAGAN is a multiple-alignment version of LAGAN and has been extensively used. The genomes are progressively aligned, then individual genomes are removed iteratively and realigned to the consensus until no improvement is found. A SHUFFLE-LAGAN version has also been developed, which can additionally find various rearrangements of sequence transversions, deletions and shiftings of DNA segments from one location to another.

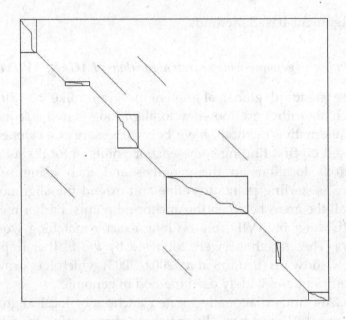

Figure 4-1. LAGAN genome alignment principle. The diagonal initial matching segments (straight lines) are joined together by a slower global alignment method (small boxes).

Though MLAGAN and similar other programs like MAVID, MAUVE, MULTIZ and TBA are popular and successful, a recent analysis indicates that their assumptions in making the alignments do not reflect well evolutionary processes, so that errors remain in the multiple alignments (Löytynoja & Goldman 2008).

Such alignment errors can be addressed by using more information about the genomes to be aligned. This has been partially achieved by the latest genome aligner, ORTHEUS, based on PECAN aligner (Paten *et al.* 2009). Traditional **phylogenetics** algorithms have focused on optimizing short sequence alignments and building trees for evolution of genes, but the improved methods in PECAN have been developed for **phylogenomics** based on genome-scale multiple alignments. PECAN software improves alignments by taking phylogeny data into account when doing a progressive alignment of several genomes and thus improves the both the quality of phylogenetic trees and the quality of alignments.

Pecan-based multiple alignments were used in ORTHEUS to reconstruct a mammalian ancestor genome alignment (Paten *et al.* 2008). In this system the optimal alignments are arrived at by combined information of matching nucleotides and phylogenetic relationships so that the most likely locations of gaps and insertions are selected (see Figure 4-2 for an example). The benefit of using multi-species alignments has been convincingly shown to improve the sensitivity of finding gene homologs and delineating the correct borders of exons. Even remotely related genomes bring a real benefit in comparative genomics, especially for the detection of short protein coding regions (Liu *et al.* 2008).

While most genome-wide search algorithms rely on just text-string based matches, some integrate additional information, *e.g.* structural information for selecting the significant matches. INFERNAL ("INFERence of RNA ALignment") algorithm searches DNA sequence databases for transcribed RNAs with a combination of RNA folding structure and sequence matches (Eddy 2006). It is based on profile stochastic context-free grammars called Covariance Models (CMs), which combine scores of sequence similarity and RNA secondary structure similarity.

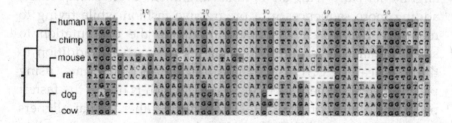

Figure 4-2. ORTHEUS multiple alignment showing an insertion in mouse and rat sequences on the left and a short 2-base deletion in dog sequence in the middle. Adapted from *Paten et al.* (2008). Reproduced with permission from CSHL Press.

This combination of features makes the search more specific with less false positives and can detect also more remote homologs, which have conserved structure but an already quite divergent RNA sequence. As in proteins, for many conserved RNAs the structure can be more conserved than the sequence. This is especially so for ribozymes and also for microRNAs and ribosomal RNAs.

Another major recent RNA alignment tool aided by structural similarities is RNAMOTIF which searches a database for RNA sequences that match a "motif" describing secondary structure interactions (Macke *et al.* 2001). The number and types of algorithms for RNA structural comparisons and discovery of non-coding RNAs de novo from raw genome sequences is increasing rapidly. Such algorithms are catalogued and reviewed in George & Tenenbaum (2009), which lists 15 RNA databases, 23 RNA structural search tools and 40 consensus alignment generators for aligned or unaligned sequences. RNA structural alignment is discussed further in chapter 5.2.

Returning to the multiple genome assembly problems caused by the recent flood of sequenced genomes, one of the most pressing challenges is that the current algorithms are still too time-consuming to be used to align thousands of microbial genomes and hundreds of eukaryotic genomes in a fast and routine fashion. As we have seen above, most multiple sequence alignments start with an optimal alignment of two sequences, adding more sequences in a progressive fashion while trying to minimize the search space and to divide the computation to smaller manageable units requiring less memory. In addition to the classical CLUSTALW algorithm, newer programs using progressive alignment include MAFFT, which is one of the fastest programs of this kind (Katoh & Toh 2008). Other similar aligners include PSAlign and T-Coffee, which however do not scale well to the more demanding large or multiple genome alignments.

To meet the need for speed, new faster methods are being developed, *e.g.* a grammar-based distance measure using Liv-Zempel compressibility has recently been used to align genome

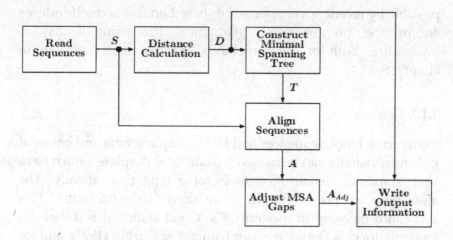

Figure 4-3. Multiple sequence alignment using compressibility distances. From Russell *et al.* (2008). Reproduced by Open Access policy of the publisher BioMed Central.

sequences. One such method resulted in a 4 to 6 fold speedup compared to MAFFT version 6 (Russell *et al.* 2008). Figure 4-3 shows the general flow of the method. Initially the distances between the sequences to be aligned are calculated using comparison of the dictionary sizes of repetitive words extracted by the Liv-Zempel compressor. After the distance matrix of NxN distances for N genomes is known, a minimal spanning tree is obtained to determine the order in which the different genomes are gradually added to the alignment, starting with the most similar genomes. After alignment of all sequences, a post-processing step further improves the gap locations to enhance the alignment quality. The computational complexity is $O(N^2 + L^2N)$, where L is sequence length and N is the number of genomes.

Further improvement of the alignment quality still appears necessary, but the speedup is already significant, and the obtained multiple alignments could be post-processed further by another algorithm. Full testing with longer genomes is still needed though. Ultimately the new post-genomic age still needs orders of magnitude faster multiple alignment technologies, which may be possible by recent novel approaches based on full-text self-indexes and/or the use of GPUs (Graphical Processing Units) in processing. Both of these methods are discussed later on in chapter 9.6.

4.1.3 *Dotplots*

Comparing DNA sequences and even complete viral and bacterial genomes visually can be made by traditional dotplots, which have been standard visualization tools for a long time already. The algorithms are nowadays also more effective due to faster CPUs and more memory in modern PCs. Good standard software for Unix platform is Dotter[2] by Sonnhammer & Durbin (1995), and for PCs the platform independent Java software JDotter (Brodie *et al.* 2004) works well for even large sequences, provided enough RAM is available. JDotter allows easy manipulation of the stringency of word matches and grey-scaling of the similarities to show the essential matches as required. A sample view of a short sequence alignment is shown in Figure 4-4.

The generation of dotplots to compare large eukaryotic genomes to each other can be slow. However, the use of k-mer suffix trees can speed up the necessary calculations markedly, as shown by Krumsiek *et al.* (2007), who developed a fast dotplotter called GEPARD. This illustrates that genome comparisons using k-mers are very useful in many contexts, as also discussed later on in chapter 4.2.1.

S-plots are another variation of the k-mer method to visually analyze genomes, for the purpose of finding areas which have unusual k-mer frequency distributions either within one genome or between two or more genomes. Developed by Putonti *et al.* (2006), this method appears to be undervalued, but has recently been reviewed favourably in Zhaxybayeva (2009). It appears that local genome segment frequencies of k-mers is a sensitive indicator of the evolutionary history of the genome segment, in a manner similar to comparing whole genomes to each other based on k-mer frequency distributions (see chapter 4.2.1).

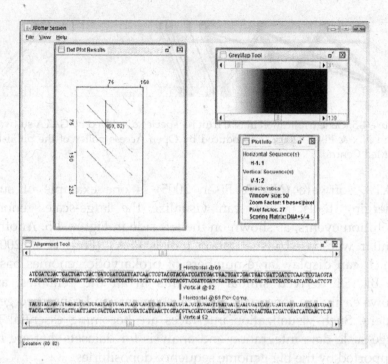

Figure 4-4. The visualization window of JDotter software. Cross-hairs in the dotplot show the location of the nucleotide alignment window below for a fictitious pair of sequences.

4.1.4 *Visualization of genome comparisons*

Good visualization of pairwise genome alignments is essential for fast and effective human eye initial analysis of genome pairs, especially for large-scale rearrangements, deletions and insertions. The large genome sequence depositories NCBI and EBI have sophisticated genome comparison displays for web browsers. These are, however, maintained and updated for the fully sequenced and annotated genomes only, with preselected and pre-computed pair-wise genome comparison visualizations. Thus tools for *ad hoc* comparisons of new genomes to each other or to known genomes are needed for either web-based or local PC based genome analysis.

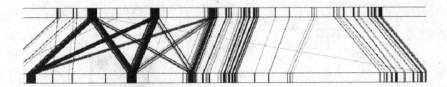

Figure 4-5. Gene triplication in two fruit fly species displayed by GATA software. From Nix & Eisen (2005). Reproduced by Open Access policy of the publisher BioMed Central.

GATA software (Nix & Eisen 2005) is one example of such emerging tools, which can visualize the large-scale genome evolution events, as shown on the left side in Figure 4-5. Another similar web-based visualization tool is PSAT (Fong *et al.* 2008), which can display large numbers of prokaryotic genomes based on BLAST derived similar segments in all the genomes, and shows easily the conserved genome segments and conserved gene orders. Such visualization interfaces are becoming essential for large-scale data integration and analysis for datasets that are not supported by the big genome sequence depositories.

For visualizing overlapping genes (both sense-sense and sense-antisense gene pairs) on genome scale, a recent system called EVOG (Cho *et al.* 2008) looks a useful tool, too. It mines the orthologs of specific genes from various genomes and calculates a similarity measure based on the proximity and order of related genes in each genome to derive a phylogenetic distance between the gene sets, *e.g.* a multi-gene family. The example shown in Figure 4-6 shows a visualization of a multigene family from three mammalian species and shows clearly the difference between mouse compared to human and chimpanzee for the HOX gene family.

4.1.5 *Global motif maps*

Full genomic scale DNA or protein motif annotations are still very new, due to the large computational effort needed to process all

the data. Also, only recently sufficient amounts of related
genomes have become accessible for comparison to enable solid
analysis of evolutionarily conserved motifs in a genomic scale
and concomitant calculations of sensitivity, specificity and false
discovery rates.

In the large ENCODE (Encyclopedia Of DNA Elements)
project conserved motif analyses were first done on a subset of
human genome data (30 megabases, about 1% of the genome),
which relied on a combination of traditional motif-finding
algorithms and manual analysis and annotation of conserved
motifs for a large amount of both sequencing data of transcribed
mRNAs and various methods to study chromatin structure, DNA
methylation and histone methylation (Encode Project Consortium
et al. 2007).

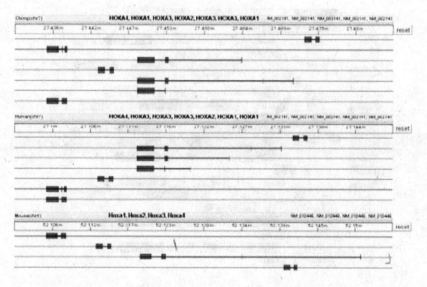

Figure 4-6. EVOG genomic comparison display of a multigene family from
three species. Chimpanzee and human (top and middle) have several copies of
HOXA family genes, but mouse (bottom) only has one copy of each type of
HOXA gene. From Cho *et al.* (2008). Reproduced by permission from IEEE,
© 2008 IEEE.

Only very recently has the first global map, called MotifMap, of conserved DNA motifs been created for the human genome (Xie *et al.* 2009). The MotifMap contains 1.5 million putative regulatory sites for 380 known regulatory motifs. Many of the motif sites coincide with known single nucleotide polymorphisms (SNPs), and may have phenotypic effects. This large dataset already gives an overabundance of testable hypotheses for experimentalists to verify the true function of these putative regulatory motifs. However, such large datasets also demand easier access and significance filtering of the data by the visualisation software.

Such visualizations are ideally web-based. One promising web display system in this direction is the recent gene ortholog comparison interface GeConT2 shown in Figure 4-7. GeConT2 integrates phylogenetically related gene complexes from COG

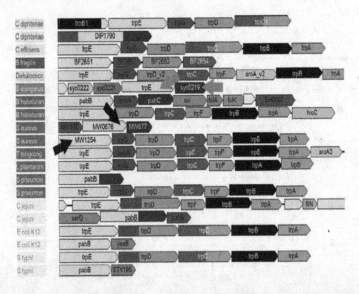

Figure 4-7. An example of the operon organization annotation display of GeConT2 (Martinez-Guerrero *et al.* 2008), which shows the orderings of metabolically related genes in different microbes. Arrows point to genes for which a function could be assigned based on the operon similarity to other operons and their annotations. Note that in the web interface the related genes are colour-coded to aid visualization. Reproduced by permission from Oxford University Press.

orthologs database with common sequence motifs of PFAM and functional annotations to metabolic data from KEGG (Martinez-Guerrero *et al.* 2008). The visualization shows the organization of operons (groupings of metabolically related genes in specific orderings) which can aid in assigning putative functions to unannotated genes.

Hu *et al.* (2006) developed an ensemble method called EMD using combined predictions of five different algorithms (AlignACE, Bioprospector, MDScan, MEME and MotifSampler), resulting in better prediction and visualization than any of the tools alone. It is clear that ensemble methods can in general outperform individual tools, provided the information used by the pooled tools is not completely overlapping.

In addition to using several string-based methods in an ensemble, one can also add other kinds of information to support predictions or   filter out false positives, for example using phylogenetic footprinting (Carmack *et al.* 2007), or previously known structures of transcription factors corresponding to DNA binding sites for predicting binding sites of novel related transcription factors (Kaplan *et al.* 2005).

One of the latest new ensemble motif tools is CompMoby (Chaivorapol *et al.* 2008), which analyses motif candidates in three groups: conserved alignable motifs, conserved, but not alignable and species-specific. The motifs from the three groups are then merged, analysed for significance and displayed.

In general, using any additional relevant data helps to enhance performance. In modern biological research there is almost always some other data available relevant to the problem, at minimum, genome or gene sequences from some related organisms or gene families.

4.2 Alignmentless methods

Alignmentless methods include several different types of approaches: word string or k-mer frequency based, sequence compressibility based, and sequence vectorization based (*e.g.* DNA walks in 2D). At genome scale, presence and absence of genes and gene sets can also be used to compare genome similarity, but require prior annotation to that effect. Of these approaches, the k-mer based approaches appear most promising, as already discussed in several earlier chapters. Recently for example, Lu *et al.* (2008) used 6-mers in phylogenetic analysis and developed a method to select the optimum length of k-mer for analysis. Therefore k-mer methods for global genome analysis are discussed at length below, followed by sequence vectorization of DNA into 2D DNA walks and 2D genome displays.

4.2.1 *K-mer based methods*

As in short motif finding algorithms utilizing informative over- and under-abundant k-mers (discussed in previous chapter 3.1.3), full genome sequences can also be compared using k-mers. This has a clear computational efficiency advantage over the tedious alignment of whole sequences or large numbers of matching genes or proteins. Another important advantage is that genomes which have little similarity and few alignable regions due to large evolutionary distance can be compared. The large variety of k-mer approaches that have been tried are reviewed by Vinga & Almeida (2003). Here only a selection of representative and promising methods are discussed.

The k-mers of oligonucleotides present in the whole genome can be counted for their frequencies, thus compressing the genome sequence information into a kind of fingerprint giving the relative frequencies of k-mer sequences. For DNA these would be A, T, C, G, AA, AT, AC, AG, TA, TT, TC, TC, ... AAA, AAT, AAC,... *etc.* In the case of DNA, each k-mer of length k will have $N = k^4$ possible oligonucleotide strings, and in the case of proteins $N = k^{20}$ possible oligopeptide strings. Under random occurrence assumption as a baseline, for a sequence of length L, the frequencies of each nucleotide or amino acid $f(a^1, a^2, a^3,...)$

determine the probability of occurrence of a specific k-mer $p(a^1, a^2, a^3,...)$ as follows:

$$p(a^1, a^2, a^3,...) = f(a^1, a^2, a^3,...) / (L-K+1),$$

where L is length of the sequence and K is the size of the k-mer being considered.

Blaisdell (1986) was among the first to use of such k-mer counts to compare DNA sequences and numerous variants of applying the principle have been published for different bioinformatics analyses. One of the most useful application areas for k-mer analysis is phylogenetics. The k-mer frequency profiles method can be used for full genomes, which would be too big for standard multiple alignment based phylogenetic methods. Both full DNA genome and concatenated proteomes can be used as input data. An early example of this approach is Qi *et al.* (2004), who produced a successful prokaryotic phylogeny based on proteome k-mer comparisons.

A more sophisticated method in the same vein for genomic DNAs was developed by Sims *et al.* (2009). A subset of k-mers 10–14 bases long was selected based on the length of the genomes to be compared, because k-mers longer than 14 gave less reliable phylogenies and k-mers shorter than 10 were also less informative. The distance between two vectors of k-mer frequencies was calculated using Jensen-Shannon divergence measure (Lin 1991). The resulting species trees agreed well with known standard phylogenies of mammals. A nice feature of their method is that genomes of widely different sizes (more than four-fold difference) can be made more comparable by dividing the larger genome to blocks of the size of the smaller genome prior to k-mer frequency profiling.

The input data in Sims *et al.* (2008) data consisted of concatenated introns only, so the method assumes that such intron annotation data is available. Whether the method could be used to compare full genomes without exclusion of non-intronic parts (*i.e.* non-coding regions) is currently unknown. It would be interesting to know how much more (or less) phylogenetic information would the k-mers from the non-intronic parts of the genome give for the phylogeny tree construction, and whether the method would scale up to use complete genomes in the analysis. The advantage of the method of Sims *et al.* is that even genomes in which it is difficult to find homologous protein coding regions between genomes can still be compared. The workflow of their method is shown in Figure 4-8.

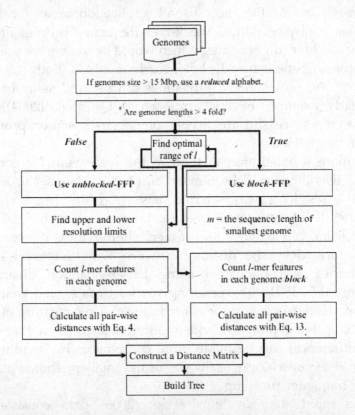

Figure 4-8. The k-mer frequency profile based method workflow of Sims *et al.* (2009). FFP = feature frequency profile (= k-mer frequency profile). Reproduced with permission from National Academy of Sciences, U.S.A.

An interesting application of k-mer analysis is finding the **missing** k-mers of a genome, and comparing genomes on the basis of common/different missing k-mers (Breland *et al.* 2008). This in effect condenses the information from each genome to a very small and specific subset of k-mers which appears to be a sensitive indicator of phylogenetic relatedness. This new method of using missing k-mers would be applicable to large datasets too, because recently a very fast method to calculate missing k-mers (called "unwords") has been developed (Herold *et al.* 2008). One should be wary, however, of attaching any special biological significance to the missing k-mers *per se*, because at least some of the bias in k-mers from random expectation seems to result from a general higher CpG mutation rate compared to other dinucleotides (Acquisti *et al.* 2007), rather than any negative selection acting to suppress specific k-mers

In a similar vein, Haubold *et al.* (2008) used shortest **absent k-mers** between two genomes to build phylogenetic trees. The method appears to be both faster and more sensitive than the traditional ClustalW phylogeny algorithm which is commonly used as a baseline method to build phylogenies. Further validation of this method is necessary, however, especially as it appears to be limited to comparisons between genomes that are not too far diverged from each other.

K-mer analysis appears useful also for filtering out known/undesired sequence data from metagenomic sequence datasets. Liu *et al.* (2008) have shown that it is possible to compute the "missing" 14-mers from the human genome with 2 or more mismatches to any human sequence. Such a set of sequences is only 2.5 million oligomers, which could be put on a chip to detect any bacteria or other non-human sequence (or deletions/insertions/transversions in *e.g.* cancer).

This could be very useful for diagnosis of human diseases, detection of bacteria and viruses and even for human gut metagenome analysis (to filter out contaminating human sequences from RNA or DNA samples from gut). The brute force alternative to getting the same data would be full sequencing of the tissue or environmental sample and subtraction of the human genome from the sequence reads, which is obviously much more expensive. Thus k-mer analysis can be used as a quick method of finding out, whether a specific sequence read is something novel and interesting, compared to the standard dataset, *e.g.* human genome from a normal healthy person without viral or bacterial diseases.

The enumeration of missing words may be useful for other genomic scale comparison tasks as well. One such area is characterization and comparison of repetitive sequences within and between genomes. Several currently used different repeat-finding algorithms (like RepeatMasker, ReCon, RepeatFinder and RepeatGluer) give markedly different, non-overlapping results (see the recent review of Saha *et al.* 2008). What is still missing in the repeat discovery algorithms is a generally accepted objective criterion of a repeat and a universal method of determining the existence and number of mutated repeats. This is a complex problem, because a specific background model for the evolution and mutations changing/erasing the repeats appears necessary.

One possible direction of future research in this area could be a linguistic analysis approach, or finding a suitable background "random" expectation model for repeats occurring by chance and looking for over-abundant repeat candidates with flexible similarity criteria. For example, suffix trees can be used to calculate the shortest missing words in a sequence to derive a measure of repetitiveness, which is fast to calculate and does not rely on alignments (Haubold & Wiehe 2006). This approach could be used by sliding windows of various sizes along the genome to find out, if repetitiveness would be a good predictor of introns, promoters *etc.*

The k-mer approach for large scale phylogeny and other genome and proteome analyses appears to have been undervalued in the past, but should now be considered a promising approach, as indicated by the recent publications discussed here. More extensive comparison and integration of these methods to traditional phylogenetic methods is to be encouraged. These methods would seem to be most useful for comparing highly divergent taxonomic groups, in which intermediate taxons showing alignable DNA or protein segments are largely lacking, for example viruses and single-celled marine eukaryotic protists (see Epstein and Lopez-Garcia 2008). This is because it appears that the composition of short k-mer motifs is more resistant to evolutionary change than the longer DNA or protein motifs which can evolve quite quickly.

New developments in the k-mer approach keep appearing. Recently a generalized form of computing distances between sequences by exhaustive k-mer composition enumeration has been presented (Apostolico & Denas 2008), which is also claimed to be very fast and to scale linearly with the length of the sequence for which the frequencies of different k-mers are calculated for. The vector of discovered k-mer frequencies can then be compared to estimate the distance (= sequence difference) between two sequences. In the same vein, Liu *et al.* (2008) reported use of n-nary k-mer profiles for remote protein homolog detection. The k-mer data was processed by latent semantic indexing and SVM and their system could detect remote protein homologs better than PSI-BLAST. Yoon *et al.* (2008) analysed over-represented k-mers in gene UTRs to pinpoint areas involved in regulatory protein binding and such areas could in fact be searched elsewhere in the genome by the discovered k-mers.

Further testing, scale-up and adaptation of these promising approaches utilizing k-mers is necessary for wider use. In conclusion, the k-mer analysis may prove to be very useful tool in the post-genomic era of extremely high throughput sequencing data flood from a variety of genome projects unravelling previously unknown sequence diversity in the biosphere from completely unknown organisms.

4.2.2 *Average common substring and compressibility based methods*

Apart from k-mer signatures, other alignmentless methods include average common substring (ACS) comparison by compression based distances and various sequence coded information entropy based methods. In the average common substring (ACS) approach (Burstein *et al.* 2005, Ulitsky *et al.* 2006) the ACS represents the average length of the longest substrings starting at any sequence position that are common between two sequences (genomes). The longer this sequence is, the more similar the two sequences are. The inverse of this value can be used as a distance measure between genomes, and phylogenies can then be built using all such pairwise genome distances. This method also scales to linear time O(N) with the length of the sequences to be compared.

The ACS method produced promisingly accurate phylogenetic trees for microbial and viral evolution using both genomes and proteomes, with fast computation times. It correctly partitioned for example prokaryotes from archaeal microbes as well as reproduced major eukaryotic groupings. Analysis of 16S ribosomal RNA sequences also resulted in trees corresponding to traditional phylogenetic trees based on multiple alignments, and analysis of several viral sequence datasets produced similarily consistent results (see Figure 4-9).

Another way to look at similarity of long genome strings is to see, how well one genome can be compressed, given the knowledge of another genome. The larger the compression is, the larger is the similarity between the two genomes. This compressibility is known as the **Kolmogorov complexity**, and methods based on sequence compressibility have been tried in several contexts for biosequence analysis. This method was used by *Li et al.* (2001) to build mitochondrial genome phylogenies and by Kuryshev *et al.* (2008) to classify primate retroviruses.

For the compression, the compression algorithm Lempel-Ziv is most commonly used, but any efficient compression method could in principle be chosen. In a recent paper, Kocsor *et al.* (2006) showed that Lempel-Ziv compression based distance measures between proteins could cluster distantly related proteins especially well. They also found that compressibility distance measures combined with BLAST could give better results than

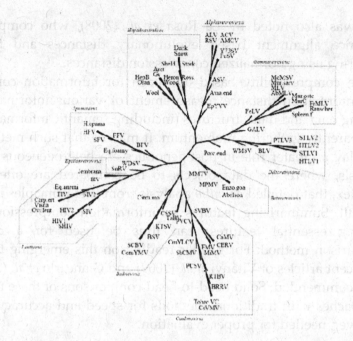

Figure 4-9. The phylogeny tree of retroviruses based on average common substring (ACS) distance method. Reproduced from Burstein *et al.* (2005) with permission from IEEE, © 2006 IEEE.

mere Smith–Waterman similarity search. This obviously comes with added complexity and computational cost, so further work is necessary to establish whether the benefits exceed the costs.

Useful reviews of compression based algorithms are those of Kertesz-Farkas *et al.* (2009) and Mantaci *et al.* (2008). They found that compressibility distances may be most useful in analyzing protein sequences with domain shuffling, but that otherwise these methods may not give a marked advantage in accuracy of the result over traditional sub-string based BLAST and Smith–Waterman algorithms. However, a computational speed advantage may be realized. Also, as indicated above, combining the compressibility information with other methods may improve similarity search, as the information retrieved by the compressibility-based methods is a somewhat different aspect of similarity than the BLAST or Smith–Waterman similarity scores.

This was also noted by La Rosa *et al.* (2008), who compared sequence alignment based evolutionary distances and SOM clustering using normalized compression distances.

The compressibility based approach for information content and information distance measurement of various information-bearing data and data structures (including semantic information) is apparently gathering momentum. It may be that such methods can play a greater role in post-genomics era heterogeneous data analysis, where the data objects to be compared are often so complex, that detailed *ab initio* analysis from first principles is too difficult. Summarizing features by information compression (or filtering essential features) can thus be used for a quick comparison method. For further reading on this emerging topic, the recent articles of Vitányi *et al.* (2009) and Giancarlo *et al.* (2009) are recommended. Solid head-to-head comparisons of these novel approaches with traditional methods for speed and accuracy are, however, needed for proper evaluation.

4.2.3 *2D portraits of genomes*

Visual processing by the human eye is extremely powerful and still surpasses the abilities of computers and artificial intelligence in general purpose pattern recognition and in 2D and 3D scene evaluation for salient aspects of the spatial information. Therefore it is attractive to transform genomic and proteomic data into 2D "portraits" of genomes for exploratory analysis by the human eye. Many different features from genome sequences can be represented in two dimensions for such visual analysis. Simple dot-plots for comparing similarities between two genomes in subchapter 4.1.3 are a simple example. Such dot-plots can also be used to show *e.g.* repetitive regions in the genome, provided a library of repeats is at hand for the genome in question (Tóth *et al.* 2006). However, this subchapter discusses some other types of interesting visualization methods, which all try to compress the genomic information to a 2D space various ways, either in the form of curves or grey-scaled or coloured pixels.

4.2.3.1 *Fractal genome displays*

For large-scale sequence data compact visualization techniques
are needed. One way to approach this problem is to map the
sequence or its oligomers into 2D space. This can be achieved by
fractal representation of DNA Sequences by iterated function
system (IFS) and Chaos Game formalism. In the iterated function
system DNA sequences or oligomer are mapped iteratively to
unique points in a 2D space. All oligomers of DNA with a fixed
size can be mapped to a 2D space with 2Nx2N elements. The
principle of the method is shown in Figure 4-10. Such
representations were already proposed in the early 1990s for short

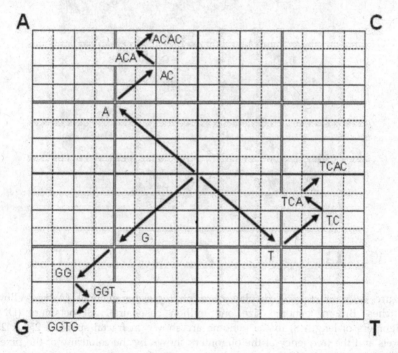

Figure 4-10. The principle of chaos game representation of DNA oligomers in 2D
fractal space. Starting from the middle point, a step is taken into the four corner
directions according to the successive letters in the DNA sequence. At each step,
the step size is halved, so that an arbitrarily long oligomer can be coded into a
finite 2D space. From Havukkala *et al.* (2006). Reproduced with permission from
IEEE, © 2006 IEEE.

DNA sequences of genes (Jeffrey 1990) and individual protein sequences (Fiser *et al.* 1994), but have not become widely used. Later on, the method has been used to visualize whole genomes as well, when personal computers became more powerful (Hao *et al.* 2000, Almeida *et al.* 2000).

In a fractal representation space similar subsequences are located close to each other. The end positions of all the oligomers in a gene or genome are marked on the grid, and the frequency of

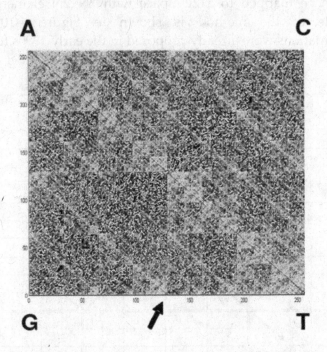

Figure 4-11. Fractal 2D display of a phytoplasma genome (Aster Yellows Witches' Broom, genome size one million basepairs). All octamers (DNA oligomers of length 8) in the genome are shown in fractal space of 256 x 256 pixels and the frequency of the oligomers shown by the colouring of the pixels. Light areas show high frequency oligomers. The light coloured diagonal from A to T indicates the large proportion of AT-rich sequences and arrow points to a cluster of abundant GT oligomers known to be enriched in this genome. Adapted from Havukkala *et al.* 2006. Reproduced with permission from IEEE, © 2006 IEEE.

each oligomer can be displayed by greyscale or colour scale pixels. An example of such oligomer frequency analysis is shown in Figure 4-11. The display shows the clusters of abundant overlapping oligomers in the genome. For example, it reveals at a glance the AT-rich nature of the genome as a whole, the similar abundance of A-repeats and T-repeats and the enrichment of GT repeats. The GT repeats may be related to host-pathogen interactions in this plant intracellular parasitic microbe, as has been suggested for other microbes (Aras *et al.* 1999). In the *Neisseria* microbe octamer repeats are also enriched and may be related to genome evolution during parasitism (Saunders *et al.* 2000).

Thus fractal analysis can quickly identify salient features in oligomer representation in whole genomes and even lead to new interesting hypotheses. Comparison between sets of genomes with this kind of method could also be illuminating to pinpoint the changes in the composition of specific oligomers between species during evolution.

If the oligomer lengths are extended to size 20–25, this fractal representation could map all unique single-copy sequences in a eukaryotic genome. Then genomes of different species could be easily compared visually at a glance. Such visualizations could even be compared by image processing methods to locate subtle differences between closely related genomes without the need to do exhaustive BLAST based alignment of whole genome against whole genome. This is an example of moving a genome comparison problem from aligned linear string analysis to bitmap image analysis domain (see the later subchapter 4.3 for another example of this kind approach for RNA structure comparisons via bitmap image representations). In general, fractal representation of biosequence data could be advantageously analysed by modern image processing methods.

The fractal type chaos game representation is under continuous development, and has been used in another format for calculating and comparing global entropies in genome sequences (Vinga & Almeida 2004, 2007). Local motif comparison of smaller DNA segments is also possible and allows for scale-free analysis of local motif statistical significance by information content as measured by entropy (Fernandes *et al.* 2009). The coordinates in the fractal space for genomic sequences have been transformed to time series and analysed with time series ARFIMA model, which showed long range correlations in the sequences (Gao & Xu 2009).

Recently the chaos game related Renyi entropy measures were used to use successive DNA letter transition probabilities in linear walk along genes to detect DNA binding site motifs (Perera *et al.* 2008). Thus the fractal space mapping of DNA sequences and Renyi entropy measure use in bioinformatics data analysis remain intriguing methods that may offer advantages in large genome visualizations. This literature appears thus worth following.

4.2.3.2 *DNA walks*

DNA sequences have long been visualized as various walks in 2D space (*e.g.* H-curve of Hamori *et al.* 1983). 2D walks of DNA sequences were used for visual comparison of sequences, e.g. Yau *et al.* (2003) method produced time-series like displays of DNA sequences (see Figure 4-12), and more sophisticated versions of this approach have been recently developed by Zhang (2009). After transforming the sequence text string to a curve, a variety of curve analysis algorithms can then be used to analyse and compare curves of different DNA sequences.

DNAs can also be coded into 3D walks and distances between curves calculated for phylogenetic comparisons of genomes (*e.g.* Luo & Zhang 2007 and Herisson *et al.* 2007), and such methods may thus move beyond mere visualization. In fact a large variety of various 2D, 3D and even 4D spatial curve codings have been suggested, but so far no large-scale analysis with these methods has been made to compare them with traditional methods to show a convincing advantage.

In any case, moving the linear text string to a sequence of numerical vector values does make available a new set of analysis

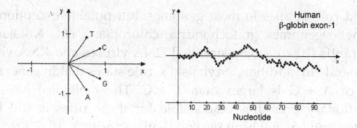

Figure 4-12. 2D DNA walk method of Yau *et al.* (2007). Left: different nucleotides are coded as steps in different directions in a 2D plane: right: an example track of one gene. Reproduced with permission from Oxford University Press.

methods that may uncover new aspects from the DNAs not accessible by traditional string analysis methods. One interesting method to analyse statistical properties of genome sequences is to use an econometric ARMA (AutoRegressive Moving Average) modelling for a 2D DNA walk (Zielinski *et al.* 2008). The ARMA analysis showed that genome coding areas appear to have more random nature than the non-coding regions, but the evolutionary pressures responsible for this are unknown. Econometric modelling techniques might reveal new non-obvious aspects of genome structure statistics when utilized in DNA sequence analysis.

Similar methods have been used also for proteins. A recent example is that of Gu *et al.* (2007) who encoded physical properties of each amino acid to a 3D walk for each protein for easy visual comparison of proteins.

4.2.3.3 *Visualizing Chargaff's rule: nucleotide skews*

According to the famous Chargaff's first rule, in double-stranded DNA the percentage of A and T is identical, as well as the percentage of G and C, according the base-pairing of A with T and G with C. More interesting from genome analysis point of view is Chargaff's second rule that states on average the same rule also holds for single-stranded DNAs (Forsdyke & Mortimer 2000). The second rule applies to most genomes, but notable exceptions are organelle genomes (mitochondria, chloroplasts *etc.*, Nikolaou & Almirantis (2006), single-stranded DNA viruses and RNA viruses in general. In addition, Szybalski's rule states that in general the sum of A + G is larger than T + C. The biological basis and evolutionary pressures responsible for these rules is still being debated and has not been resolved satisfactorily.

Local variations to Chargaff's second rule in genomes of different organisms are known to occur. One possible explanation for the second rule is inversions and inverted transpositions in the genome during evolution, which evens out the proportions of nucleotides on both strands over long time periods (Albrecht-Buehler 2006). The most recent and attractive hypothesis is that specific oligomers are conserved in their frequency in various genomes since ancient times as a relic from the primordial genome which originally was repetitive (Zhang & Huang 2008).

Whatever the explanation of Chargaff's second rule, local variations in AT ratio and GC ratio are of intrinsic interest, because they may be useful in genome annotation and in finding interesting regions in the genome. The nucleotide skews AT skew $(A-T)/(A+T)$ and CG skew $(C-G)/(G+C)$ are known to vary significantly in local genome scale, but the reasons for the origin and maintenance of these variations is not known. The base skews have been shown to correlate with various functional domains in genomes, including gene rich areas, transcribed regions, origin of replication and so on, both in prokaryotes and eukaryotes (Bell & Forsdyke 1999, Fujimori *et al.* 2005, Majewski 2003 and Touchon *et al.* 2005).

The nucleotide skews have normally been analyzed by using cumulative skew diagrams, *e.g.* CG skew cumulative curves (Grigoriev 1998). Simultaneous plotting of cumulative total skew as sum of AT and CG skews over 1 kb non-overlapping windows was used by Touchon *et al. (2005)* and such profiles could be used to predict replication origins in mammalian genomes. Visualization software for nucleotide skews in bacterial genomes is also available (Pritchard *et al.* 2006). These algorithms, however, do not scale to large eukaryotic genomes and a more compact visualization is needed.

A compact visualization called Base Skew Double Triangle (BSDT) was developed by Deng *et al.* (2007), which also shows usefully the AT and CG skew correlation by symmetry of the two skews in different genomic scales (Figure 4-13). Deng *et al.* found a significantly higher correlation between AT skew and GC skew in birds compared to other vertebrates, which remains unexplained so far. For longer genome segments (over 20 kb), the correlation between AT and CG skew is about 90% in birds, while in other vertebrates it varies from 60 to 80%. Oligomer analysis in various vertebrate species of the most conserved segments (*e.g.* haplotype blocks) along the lines of Zhang & Huang (2008) would be interesting, because oligomer conservation might explain these differences.

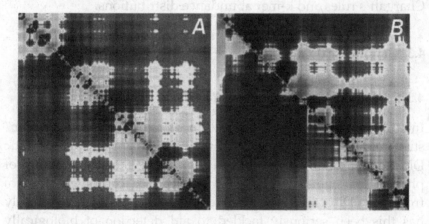

Figure 4-13. Simultaneous AT and GC nucleotide skew visualization byBase Skew Double Triangle (BSDT). AT skew: bottom left triangle, GC skew: top right. Left: chicken chromosome 5; right: pufferfish chromosome 3. Higher correlations are in darker colour. The correlation of AT and GC skews is strikingly symmetric in the chicken at various scales and the local correlation shows as a dark diagonal from top left to bottom right. From Deng *et al.* (2007). Reproduced by Creative Commons Attribution License from an Open Access article.

A similar enigma is the local variation in GC percentages along the genomes, organized in stretches of genome with a consistent proportion of G+C, called isochores. Isochores are now considered to depend independently on both recombination rate and exon (coding sequence) density (Freudenberg *et al.* 2009). The isochore structure in various genomic scales could also be visualized in a 2D triangle form to display isochore structure simultaneously in different genome scales. Wang *et al.* (2008) developed an informative colour visualization in 2D for such purpose. K-mer distributions may have a role to play. Chor *et al.* (2009) analysed k-mer spectra from 100 genomes, and found that tetrapods have a multimodal distribution of k-mers, as opposed to lower eukaryotes and other organisms. This implies that there are at least two distinct groups of k-mers, common ones and rarer ones. Integrated isochore, nucleotide skew and k-mer spectral analysis would be most interesting and could finally solve the evolutionary history and possible links between isochores, Chargaff's rules and k-mer abundance distributions.

4.3 Genome scale non-sequence data analysis

4.3.1 *DNA physical structure based methods*

In addition to the alphabet based information for "lexicographic" string analysis (which has been the mainstream of bioinformatic DNA and protein sequence analysis so far), there is a wealth of physical parameter information that can be computed *in silico* from the DNA, RNA and protein sequences. Only recently has this been seriously tackled to aid detection of biologically relevant domains and motifs along genomes. As an example, the shortcomings of mere string and motif based analyses for detecting physical modification sites of proteins have been clearly demonstrated by the Scan-X algorithm of Schwarz *et al.* (2009). Their method used a large amount of experimental data from mass spectrometry as a starting point and also collated over-abundant known phosphorylation motifs from several organisms and used the pooled data to define search motifs to scan new sequences for phosphorylation sites. Such sites are hard to find *de novo* by traditional string based analysis. Sensitivity of the method was still quite poor (*i.e.* low proportion of true positives), around 23%, but specificity is excellent, 97%.

The large number of missed true positives due to low sensitivity is problematic, because the location of an unknown phosphorylation site could be very difficult to find *ab initio* along the genome or in the protein of interest. It appears that text-based motif analysis of the proteome sequences may not obtain enough information about the relevant features to find the true phosphorylation sites. Including physico-chemical features of the amino acid motifs into classification algorithms should reduce the number of false positives for any motif for which the relevant physical parameters are known.

A landmark paper in physical profiling along the genome by *in silico* methods is Abeel *et al.* (2008a) who successfully predicted promoters in genomic sequence using a single physical parameter of base-stacking values. In this method the base-stacking values were calculated for a 400 basepair sliding window and the resulting profile along the genome appears to reliably identify eukaryotic promoters and even transcription start sites (see Figure 4-14). The method is also very fast to calculate and needs no parameter tuning. Similarly good results across many different genomes were obtained when using local small-scale 3 basepair window averages with self-organizing maps (the ProSOM algorithm of Abeel *et al.* 2008b).

The surprising finding that using just one physical parameter (base stacking) is so powerful for domain prediction is very promising and this kind of additional information could significantly boost the accuracy of text string based *de novo* DNA motif discovery from genomic sequences. It is interesting to note, that currently only approximately 15 different computable

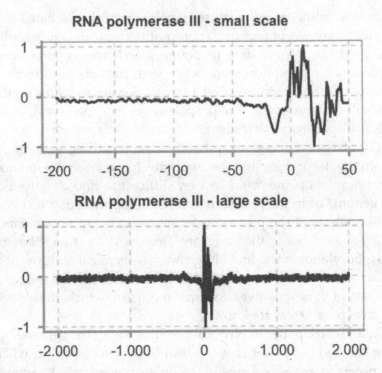

Figure 4-14. Profile of base-stacking values along the genome across transcription site start (coordinate 0 on x-axis) using a 400 basepair sliding window. From Abeel *et al.* (2008a). Reproduced by permission from CSHL Press.

parameters (obtained by non-redundant features from 115 different physical properties of dinucleotides) are necessary to represent most of the structural information in DNA (Friedel *et al.* 2009, for details, see the DiProDB website http://diprodb.fli-leibniz.de).

The physical features of DNA dinucleotides that were most informative and independent of each other were: Direction (i.e. 3′ vs. 5′), Inclination, Twist–rise (conformational), Base-stacking energy, Tilt, Shift, Propeller twist and Rise. In addition, there are many other physical parameters that can be calculated for longer windows of DNA or RNA. The parameters explored by Abeel *et al.* (2008a) included the following:

Aphilicity	Base stacking	BDNA twist
Bendability	Bending stiffness	DNA denaturation
Duplex disrupt	Duplex stability	Free energy
GC content	Nucleosome position	Propellertwist
Protein deformation	Proteindnatwist	Radical cleavage
ZDNA		

Of these parameters, some have already been explored in the genomic scale, for example DNA roll and melting enthalpy were profiled by Deyneko *et al.* (2006). Also DNA melting temperature (related to bubble formation) has been profiled along the whole human genome and was found to correlate with many known annotation features (Liu *et al.* 2007). Such bubble formation analysis has already been found to be useful for *ab initio* gene and exon finding in different genomes (Jost & Everaers 2009).

New fast methods of calculating bubble formation domains along large genomes have already been developed (Tøstesen 2008, Tøstesen *et al.* 2009). Therefore the stage is set for large-scale correlation analyses of this physical factor with known genome annotations and for discovery of novel interesting regions showing unusual patterns of bubble formation profiles in genome regions of no or little known function. There are interesting hints that the fast local dynamics of bubble formation in the adenoviral promoters are correlated with transcription factor binding (Choi *et al.* 2008).

Full genome annotation for various promoters, splice sites, transcription start site, m-RNA polyadenylation sites, exons, non-coding RNAs *etc.* might be improved dramatically by using a combination of judiciously chosen physical parameters at appropriate length scales. The appropriately weighted physical parameters could be optimized to each organism being studied and used together with the traditional motif analysis algorithms. At a minimum, combining this information with the traditional text-based motif finders will bring a clear benefit for both known motif recovery and *de novo* motif discovery. In effect, this will bring the biological motif finding effort closer to the physical chemistry domain from the current "non-biological" alphabet based text string motif analysis.

It is also interesting to note here that the GC percentage, which is often used as a classificatory feature in genome annotation, seems to be closely related to solvent-accessibility of the corresponding DNA structure (Parker *et al.* 2008). Most recently, the hydroxyl cleavage pattern (which is a measure of solvent-accessibility and affects protein-binding affinity of DNA) was used in cross-genome comparison of structural conservation (Parker *et al.* 2009). *In silico* computed binding affinities over 4 basebairs long windows along selected portions of the human genome was used to create DNA topography profiles along the genome. Similar profiles from genomes of other species were aligned to the human genome to measure structural conservation of the hydroxyl cleavage pattern along the human genome. It was found that 12% of the genome was structurally conserved, and that twice as many conserved regions could be detected by the hydroxyl cleavage pattern profiles than by traditional sequence conservation.

Furthermore, the observed patterns in the profiles correlated with many genomic features, notably also with non-coding functional elements such as promoter enhancers. Single nucleotide polymorphisms having a phenotypic effect were also found to be correlated with a change in the solvent accessibility profile. Here again, it is evident that the physical chemistry parameters are becoming very useful in genomic sequence data mining and analysis, when the genome is considered not only as a sequence of letters, but as a profile of successive physico-chemical patterns along the DNA.

Though many physical parameters of DNA can be computed *in silico* from the DNA sequence, many other aspects of the chromosomes cannot be directly calculated. These include DNA methylation and chromatin modifications for proteins wrapping around DNA. Methylation and protein modifications are now thought to control each other, with DNA methylation involved in long-term transcription control, but modulated in short term by changes in chromatin histone proteins (Cedar & Bergman 2009).

New high-throughput experimental methods are becoming available to get detailed high-resolution data of both the physical parameters and DNA and protein modifications along the genome which then have to be analysed bioinformatically. For example, new high-throughput sequencing machines in conjunction with immunoprecipitation techniques have made it possible to make global detailed maps of chromatin features which indicate actively transcribed regions (Guttman *et al.* 2009). The method revealed about 1,600 mostly unknown intergenic large transcripts (on average 20 kb long), which are evolutionarily conserved. Their function is mostly unknown, but many of them are situated close to transcription factors and genes involved in morphogenesis, cell division and immunological surveillance, that is, complex regulatory systems. Their transcription seems to be controlled also by the same transcription factor proteins as the nearby genes. The analysis of this novel class of very large non-protein coding transcripts presents interesting challenges for *in silico* bioinformatics analysis.

Proteomes (as well as single proteins) can also be analyzed in a similar manner. Physicochemical features of protein domains was used as input to a SVM classifier to cluster and detect G-protein coupled receptors (Yang *et al.* 2008) in a manner analogous with the DNA physical property profile method of Abeel *et al.* (2008a). Also for proteins, it appears that the mere text-based string analysis can be very usefully complemented with the inclusion of the physical characteristics of individual amino acids and stretches of amino acids. It is expected that this type of analysis extended into whole proteome analysis will improve remote homology detection and enable more accurate phylogenies to be constructed.

4.3.2 *Secondary structure based comparisons*

In addition to the linear physical parameters calculated along the DNA or RNA sequences, the secondary structure into which the local segments of *e.g.* long DNA molecules or short mRNA molecules fold into affects crucially their biological activity, as discussed in the introduction. Therefore high-throughput methods to calculate potential secondary structures are necessary both to characterize the folding structures *in silico* (instead of the slow and expensive crystallography experiments in the laboratory) and also to find out the potential effects of mutations in the sequence on the folding and resulting secondary structures.

In protein bioinformatics, established methods have been developed for comparing known protein 3D structures and many unknown structures can be quite reliably predicted, provided enough related known protein structures are known. *Ab initio* protein folding without known related structures is still considered a grand challenge, but is outside the scope of this book. An equally interesting and challenging task, however, is large-scale accurate analysis of folding in novel DNA and RNA sequences. This is a relatively new, but rapidly expanding research area, and is discussed at some length here.

4.3.2.1 *DNA quadruplex forming motifs*

DNA quadruplexes are four-stranded DNA helixes due to specific aligned stacking of G nucleotides. Regions which are rich in G nucleotide leading to quadruplexes are commonly found in many eukaryotic genomes, including telomeres, gene promoters, UTRs, recombination sites, and tandem repeat regions. An example of a quadruplex formation is shown in Figure 4-15. Such structures can be verified by circular dichroism spectroscopy and biochemical techniques like dimethyl sulfate footprinting.

It has been noted that in areas of high G content, alternative conformations with different stacking of G nucleotides are possible, and many different topologies for such quadruplexes

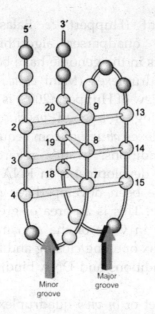

Figure 4-15. An example of the topology of G-quadruplex formation in the promoter of the human gene VEGF. The stacked G nucleotides are the horizontally linked nucleotides; numbers refer to successive nucleotides of the sequence. From Guo *et al.* 2008. Reproduced by permission from Oxford University Press and the authors.

have been shown to exist (Lane *et al.* 2008). It is thought that at least some of these conformations actually regulate the opening and closing of the promoter between 4-strand and 2-strand helixes, thus controlling the access of transcription factors to the promoter to activate the gene.

Such G-quadruplex structures in gene promoter regions of oncogenes and other crucial genes related to apoptosis are potential targets for anticancer drug development. Recently it has been shown, that G-quadruplex actually represses translation of mRNA to the human Zic-1 zinc-finger protein (Arora *et al.* 2008), and it is suspected this is a common regulation mechanism in eukaryotes. It has also been shown that human gene promoters are significantly enriched for G-rich segments capable of forming quadruplexes and that about 43% of promoters contain at least one quadruplex motif (Huppert & Balasubramanian 2008) identified using the quadparser algorithm. Recently, also recombination hotspots in the genome have been associated with the formation of quadruplexes (Mani *et al.* 2009). For further reading, the recent review of Huppert (2008) is recommended.

What is still missing is a good *in silico* modeling of the formation of quadruplexes *ab initio* from sequence data alone, in biological *in vivo* conditions. In this regard DNA folding is somewhat behind the developments in RNA folding, for which many algorithms have been developed for *de novo* predictions of reasonable accuracy. This is an area of great physicochemical complexity, especially in view of the dynamical and reversible nature of the DNA helix opening/closing and folding kinetics and the role that ionic conditions and DNA binding proteins play in these processes.

A large-scale dataset of *in vivo* quadruplex structures, *e.g.* for specific promoters, would open the door to correlating such structures to gene regulation and action of transcription factor binding. Future progress in this field requires new high-throughput structural determination methods as well as better *ab initio* quantitative prediction methods, which could calculate the probability and stability of quadruplexes as opposed to the mere presence/absence of a quadruplex site, while taking into account the metabolic/regulatory context and physiological conditions *in vivo*.

4.3.2.2 *RNA secondary structure motifs*

The theory for RNA secondary folding is well advanced, including the algorithms for *in silico* folding, and the literature is extensive. A full review of the theory of folding or competing RNA folding algorithms is not attempted here, but for a recent review of the physicochemical aspects, see Thirumalai & Hyeon (2009) and Ding *et al.* (2008), for non-coding RNA analysis in general, see Machado-Lima *et al.* (2008), and for an exhaustive listing of available RNA structure related algorithms, see Capriotti *et al.* (2008). Here we focus initially on the biophysics of RNA folding, followed by some state-of-the-art examples.

RNA folding physical chemistry has been closely studied by molecular dynamics (MD) simulations and laboratory experiments and a considerable body of information exists on the physics of the process (Thirumalai 2009), which can lead to many alternative conformations of the folding RNA even in stable conditions. The kinetic partitioning of the final competing structures is called a Boltzmann ensemble, *i.e.* a set of stable structures which are in dynamic balance, so that the intermediate structures between the local maxima are only of short duration.

The problem of RNA folding is difficult, because all four nucleotides can interact with each other, and because the folded structures are less stable than in proteins and can easily change configuration to alternative structures. Furthermore, counter-ions can modulate the folding of the RNA due to its polyelectrolyte nature. In addition, different regions of the RNA show different flexibility, dictated by sequence, which affects folding, as is known to occur in DNA as well.

It is important to note that even single nucleotide mutations are known to change the RNA folding dynamics or its most favoured configuration (Pan *et al.* 2000). Therefore accurate simulations of putative RNA structures are important for any phylogenetic studies relying on structural conservation features in classification. However, it is also important, that the partitioning into favoured configurations is affected by the initial conditions (including ionic conditions, *e.g.* cationic charge density and physical shape at the start of folding), *e.g.* during the RNA transcription or release from a RNA binding protein. The folding time also varies over many orders of magnitude between 10 and 20 bases long RNAs (Thirumalai 2009). It is not known, if the timescales of folding of longer RNAs actually are relevant in the *in vivo* conditions in the biochemical machinery of cells. A single basepair formation for RNA takes about 1 microsecond, so that folding can happen quite quickly after the synthesis of the RNA.

Quite accurate 3D structural folding simulations by sophisticated discrete molecular dynamics (DMD) methods are available, but unfortunately they are still very slow, for example a single 36 nucleotides long RNA took 5 hours to fold on a cluster of 8 Xeon 3.6 MHz CPUs using MPI. For genome scale RNA structure predictions, such sophisticated molecular dynamics simulations are yet unpractical due to the high computational demands. Therefore so far these tasks have been performed using heuristic simplified RNA folding algorithms like MFold, Vienna Package *etc.*

A recent example of a simple but effective method is that of Brameier & Wiuf (2007), who developed a linear genetic programming classifier based on simple regular expressions (similar to ProSite motifs) of predicted miRNA secondary structure bracket representations produced by sliding a window along the genome and running RNAfold of the Vienna package. An ensemble of 16 predictors worked well and seemed to outperform standard SVM type predictors. This approach combined with physical parameters of predicted miRNA locations in the genome looks a promising first pass scan method for new genomes for which no reliable training data is not yet available.

The basic problem still remains, though: the number of false positive predictions is still large, due to the huge size of genomes and the very large number of predictions by a sliding window advanced one base-pair at a time. It is as yet unknown which of the predicted positives are real potential miRNAs that have not yet evolved to full function or are not yet allowed to have a function in the metabolism by the existing control systems (most of which are still largely unknown).

Given sufficient computing resources, large-scale predictions of microRNAs using heuristic methods are already possible. Genome-scale predictions for all possible microRNAs have already been made by a sophisticated system called mirORTHO, which is based on an *ab initio* hairpin detector slid along the genome, followed by SVM filtering using homolog data, ortholog mining from other species and an SVM classifier utilizing ortholog microRNA multiple alignments. The system has been used to produce comprehensive microRNA candidate predictions for over 30 animal genomes (Gerlach *et al.* 2009). The system discovered over 150 novel eukaryotic microRNAs to add to the about 600 known previously.

Most of this kind of microRNA prediction systems rely to a high degree on suitable training sets of known verified microRNAs. The fact that several algorithms trained on one organism data perform well on another organism clearly indicates that the basic structural features of microRNAs have changed very little during the long period of eukaryotic evolution. This is in accordance with the recent theories that microRNAs evolved very early as a defence mechanism against viruses (Obbard 2009). Supporting this macroevolutionary hypothesis is the fact that promoters controlling microRNA transcription are clearly more conserved in evolution across species than the promoters of protein-coding genes (Wan *et al.* 2008). This emphasizes the importance of the regulatory system of non-coding RNAs, including microRNAs.

Recently also totally *ab initio* methods for RNA folding have improved significantly. It now appears that structural features of RNA sequence are sufficient to predict microRNAs *de novo* using hierarchical hidden Markov Models (HHMMs), reaching accuracy and sensitivity of 84% and specificity of 88% for human microRNAs and over 97% accuracy for plant microRNAs (Kadri *et al.* 2009). This makes it for the first time possible to make good *in silico* prediction of all microRNAs in most eukaryotes, which is a significant boost for the already exponentially increasing miRNA research. Other methods of similar performance include triple-SVM (Xue *et al.* 2005) and MicroPred (Batuwita & Palade 2009) which achieved 90% sensitivity and 92% specificity using 21 structural features in an SVM analysis.

Finally, an example about the practical importance of RNA folding algorithms: influenza virus RNA folding pattern appears to affect transcription activity and potentially also virus evolution. Slight changes in the RNA sequence influence crucially the 3D structure and its dynamics in the Boltzmann ensemble (Brower-Sinning *et al.* 2009). Temperature is also likely to be an important factor, as folding is strongly affected by it. Interestingly, birds have a higher temperature than mammals and are often not affected as much as humans by the influenza virus. This has important implications on the evolution and adaptation of viruses between different hosts, including obviously also insect-mediated animal viruses. This is an interesting new direction to explore, but requires easy to use, fast and reliable RNA folding algorithms and more experimental data on real life RNA structures *in vivo* in different organisms and in different temperatures to validate the predicted folding structures and their biological activity.

Ultimately, in the RNA folding research biophysics will merge with the molecular biology and bioinformatics. Hopefully Graphical Processing Unit (GPU) boosted algorithms can be used for molecular dynamics simulations effectively enough on bioinformaticians' desktop computers or cheap clusters of computers. The possible revolutionary role of GPUs in future bioinformatics is discussed in the chapter 9.8. In the meantime before the availability of GPU computing, high performance computing (HPC) is necessary for the next stage in physical chemistry based RNA informatics. Simulating in 3D the folding of all transcribed RNAs (including the non-coding RNAs) of the genome will make it possible to look for conserved folding structures and control mechanisms of RNA expression in global genomic scale.

References

Abeel, T., Saeys, Y., Bonnet, E., Rouzé, P. & Van de Peer, Y. 2008a. Generic eukaryotic core promoter prediction using structural features of DNA. *Genome Research 18(2): 310–323.*

Abeel, T., Saeys, Y., Rouzé, P. & Van de Peer, Y. 2008b. ProSOM: core promoter prediction based on unsupervised clustering of DNA physical profiles. *Bioinformatics 24(13): i24–31.*

Acquisti, C., Poste, G., Curtiss, D., Kumar, S. 2007. Nullomers: really a matter of natural selection? *PLoS ONE 2(10): e1022.*

Albrecht-Buehler, G. 2006. Asymptotically increasing compliance of genomes with Chargaff's second parity rules through inversions and inverted transpositions. *Proc. Natl. Acad. Sci. USA 103: 17828–17833*.

Almeida, J.S. *et al.* 2000. Analysis of genomic sequences by Chaos Game Representation. *Bioinformatics 17(5): 429–437*.

Apostolico, A. & Denas, O. 2008. Fast algorithms for computing sequence distances by exhaustive substring composition. *Algorithms for Molecular Biology 3: 13*.

Aras, R.A. *et al.* 1999. Extensive repetitive DNA facilitates prokaryotic genome plasticity. *Proc. Natl. Acad. Sci. USA 100(23): 13579–135784*.

Arora, A., Dutkiewicz, M., Scaria, V., Hariharan, M., Maiti, S. & Kurreck, J. 2008. Inhibition of translation in living eukaryotic cells by an RNA G-quadruplex motif. *RNA 14(7): 1290–1296*.

Batuwita, R. & Palade, V. 2009. microPred: Effective classification of pre-miRNAs for human miRNA gene prediction. *Bioinformatics 25(8): 989–995*.

Bell, S.J. & Forsdyke, D.R. 1999. Deviations from Chargaff's second parity rule correlate with direction of transcription. *J. Theor. Biol. 197(1): 63–76*.

Blaisdell, B.E. 1986. A measure of the similarity of sets of sequences not requiring sequence alignment. *Proc Natl Acad Sci USA 83: 5155–5159*.

Brameier, M. & Wiuf, C. 2007. Ab initio identification of human microRNAs based on structure motifs. *BMC Bioinformatics 8: 478*

Brodie, R., Roper, R.L. & Upton, C. 2004. JDotter: a Java interface to multiple dotplots generated by dotter. *Bioinformatics 20(2): 279–81*.

Brudno, M., Chapman, M., Göttgens, B., Batzoglou, S. & Morgenstern, B. 2002. Fast and sensitive multiple alignment of large genomic sequences. *BMC Bioinformatics 4: 66*.

Brudno, M., Do, C.B., Cooper, G.M., Kim, M.F., Davydov, E.; NISC Comparative Sequencing Program, Green, E.D., Sidow, A. & Batzoglou, S. 2003. LAGAN and Multi-LAGAN: efficient tools for large-scale multiple alignment of genomic DNA. *Genome Res. 13(4): 721–731*.

Burstein, D. *et al.* 2005. Information Theoretic Approaches to Whole Genome Phylogenies. *Lecture Notes in Computer Science 3500: 283–295*.

Capriotti, E. & Marti–Renom, A. M. 2008. Computational RNA Structure Prediction. *Current Bioinformatics 3: 32–45*.

Carmack, C.S., McCue, L.A., Newberg, L.A. & Lawrence, C.E: 2007. PhyloScan: identification of transcription factor binding sites using cross-species evidence. *Algorithms for Molecular Biology 2: 1*.

Cedar, H. & Bergman, Y. 2009. Linking DNA methylation and histone modification: patterns and paradigms. *Nat Rev Genet. 10(5): 295–304*.

Chaivorapol, C., Melton, C., Wei, G., Yeh, R.F., Ramalho-Santos, M., Blelloch, R. & Li, H. 2008. CompMoby: comparative MobyDick for detection of cis-regulatory motifs. *BMC Bioinformatics 9: 455*.

Cho, C.-Y., Lee, D.H. & Cho, H.-G. 2008 A Comparative Visualization System for the Proximity of Genes on Whole Genome Scale. *In: International Conference on Convergence and Hybrid Information Technology, ICHIT '08. Pp. 14–18*.

Choi, C.H., Rapti, Z., Gelev, V., Hacker, M.R., Alexandrov, B., Park, E.J., Park, J.S., Horikoshi, Nm, Smerzi, A., Rasmussen, K.Ø., Bishop, A.R. & Usheva, A. 2008. Profiling the thermodynamic softness of adenoviral promoters. *Biophys J. 95(2): 597–608.*

Chor, B., Horn, D., Levy, Y., Goldman, N. '& Massingham, T. 2009. Genomic DNA k-mer spectra: models and modalities. *Genome Biol. 10(10): R108.*

Deng, X., Havukkala, I. & Deng, X. 2007. Large-scale genomic 2D visualization reveals extensive GC-AT skew correlation in bird genomes. *BMC Evolutionary Biology 7: 234.*

Deyneko, I.V., Bredohl, B., Wesely, D., Kalybaeva, Y.M., Kel, A.E., Blöcker, H., Kauer, G. 2006. FeatureScan: revealing property-dependent similarity of nucleotide sequences, *Nucleic Acids Res. 34: W591–595.*

Ding, F. *et al.* 2008. Ab initio RNA folding by discrete molecular dynamics: from structure prediction to folding mechanisms. *RNA 14: 1164–1173.*

Eddy, S. R. 2006. Computational analysis of RNAs. *Cold Spring Harbor Symposia on Quantitative Biology 71: 117–128.*

ENCODE Project Consortium et al, 2007. Identification and analysis of functional elements in 1% of the human genome by the ENCODE pilot project. *Nature 447(7146): 799–816.*

Epstein, S. & Lopez-Garcia, P. 2008. Missing' protists: a molecular prospective. *Biodiversity and Conservation 17(2): 261–276.*

Fernandes, F., Freitas, A.T., Almeida, J.S. &Vinga, S. 2009. Entropic Profiler-Detection of conservation in genomes using Information Theory. *BMC Research Notes 2: 72.*

Fiser, A., Tusnady, G.E. & Simon, I. 1994. Chaos game representation of protein structures. *J. Mol. Graphics 12: 302–304.*

Fong, C., Rohmer, L., Radey, M., Wasnick, M. & Brittnacher, M.J. 2008. PSAT: a web tool to compare genomic neighborhoods of multiple prokaryotic genomes. *BMC Bioinformatics 9: 170.*

Forsdyke, D.R. & Mortimer, J.R. 2000. Chargaff's legacy. *Gene 261: 127–137.*

Freudenberg, J., Wang, M., Yang, Y. & Li, W. 2009. Partial correlation analysis indicates causal relationships between GC-content, exon density and recombination rate in the human genome. *BMC Bioinformatics 10 (Suppl 1): S66.*

Friedel, M., Nikolajewa, S., Sühnel, J., Wilhelm, T. 2009. DiProDB: a database for dinucleotide properties. *Nucleic Acids Res. 37(Database issue): D37–40.*

Fujimori, S., Washio, T., Tomita, M. 2005. GC-compositional strand bias around transcription start sites in plants and fungi. *BMC Genomics 6: 26.*

Gao, J. & Xu, Z.-Y. 2009. Chaos game representation (CGR)-walk model for DNA sequences. *Chinese Phys. B 18: 370–376.*

George, A.D., Tenenbaum, S.A. 2009. Informatic resources for identifying and annotating structural RNA motifs. *Mol Biotechnol. 41(2): 180–93.*

Gerlach, D., Kriventseva, E.V., Rahman, N., Vejnar, C.E. & Zdobnov, E.M. 2009. miROrtho: computational survey of microRNA genes. *Nucleic Acids Res. 37(Database issue): D111–7.*

Giancarlo, R., Scaturro, D. & Utro, F. 2009. Textual Data Compression In Computational Biology: A synopsis. *Bioinformatics [Epub ahead of print]*.

Grigoriev, A. 1998. Analyzing genomes with cumulative skew diagrams. *Nucleic Acids Research 26(10): 2286–2290*.

Gu, S., Poch, O., Hamann, B. & Koehl, P. 2007. A geometric representation of protein sequences. *Proceedings of the 2007 IEEE International Conference on Bioinformatics and Biomedicine, pages 135–142*.

Guo, K., Gokhale, V., Hurley, L.H. & Sun, D. 2008. Intramolecularly folded G-quadruplex and i-motif structures in the proximal promoter of the vascular endothelial growth factor gene. *Nucleic Acids Res. 36(14): 4598–4608*.

Guttman, M., Amit, I., Garber, M., French, C., Lin, M.F., Feldser, D., Huarte, M., Zuk, O., Carey, B.W., Cassady, J.P., Cabili, M.N., Jaenisch, R., Mikkelsen, T.S., Jacks, T., Hacohen, N., Bernstein, B.E., Kellis, M., Regev, A., Rinn, J.L & Lander, E.S. 2009. Chromatin signature reveals over a thousand highly conserved large non-coding RNAs in mammals. *Nature 458(7235): 223–2237*.

Hamori, E. & Ruskin, J. 1983. H curves, a novel method of representation of nucleotide series especially suited for long DNA sequences. *J. Biol. Chem. 258: 1318–1327*.

Hao, B., Lee, H. & Zhang, S. 2000. Fractals related to long DNA sequences and complete genomes. *Chaos, Solitons and Fractals 11(6) (2000) 825–836*

Haubold, B., Pierstorff, N., Möller, F. & Wiehe, T. 2005. Genome comparison without alignment using shortest unique substrings. *BMC Bioinformatics 6: 123*.

Havukkala, I., Benuskova, L., Pang, P., Jain, V., Kroon, R. & Kasabov, N. 2006. Image and fractal information processing for large-scale chemoinformatics, genomics analyses and pattern discovery. *In: Lecture Notes in Bioinformatics LNBI 4146: 163–173*.

Herisson, J., Payen, G. & Gherbi, R. 2007. A 3D pattern matching algorithm for DNA sequences. *Bioinformatics 23(6): 680–686*.

Herold, J., Kurtz, S. & Giegerich, R. 2008. Efficient computation of absent words in genomic sequences. *BMC Bioinformatics 9: 167*.

Hu, J., Yang, Y.D. & Kihara, D. 2006. EMD: an ensemble algorithm for discovering regulatory motifs in DNA sequences. *BMC Bioinformatics 7: 342*.

Huppert, J.L. 2008. Hunting G-quadruplexes. *Biochimie 90(8): 1140–1148*.

Huppert, J.L. & Balasubramanian, S. 2007. G-quadruplexes in promoters throughout the human genome. *Nucleic Acids Res. 35(2): 406–413*.

Jeffrey, H.J. 1990. Chaos game representation of gene structure. *Nucleic Acids Res. 18(8): 2163–2170*.

Jost, D. & Everaers, R. 2009. Genome wide application of DNA melting analysis. *J. Phys. Condens. Matter 21: 034108 (14pp)*.

Kadri, S., Hinman, V. & Benos, P.V. 2009. HHMMiR: efficient de novo prediction of microRNAs using hierarchical hidden Markov models. *BMC Bioinformatics 10(Suppl 1): S35*.

Kaplan, T., Friedman, N. & Margalit, H. 2005. Ab initio prediction of transcriptionfactor targets using structural knowledge. *PLoS Comput Biol. 1(1): e1.*

Katoh, K. & Toh, H. 2008. Recent developments in the MAFFT multiple sequence alignment program. *Brief Bioinform. 9(4): 286–98.*

Kertesz-Farkas, A., Kocsor, A. & Pongor, S. 2009. The Application of Data Compression-Based Distances to Biological Sequences. *In: Information Theory and Statistical Learning, Springer US, pp. 83–100.*

Kocsor, A., Kertész-Farkas, A., Kaján, L. & Pongor, S. 2006. Application of compression-based distance measures to protein sequence classification: a methodological study. *Bioinformatics 22(4): 407–12.*

Krumsiek, J., Arnold, R. & Rattei, T. 2007. Gepard: a rapid and sensitive tool for creating dotplots on genome scale. *Bioinformatics 23(8): 1026–1028.*

Kuryshev, V.Y. & Hanus, P. 2008. Compression Based Classification of Primate Endogenous Retrovirus Sequences. *In: Proc. 5th International Workshop on Computational Systems Biology (WCSB 2008), Leipzig, Germany, pp. 81–84.*

Lane, A.N., Chaires, J.B., Gray, R.D. & Trent, J.O. 2009. Stability and kinetics of G-quadruplex structures. *Nucleic Acids Res. 36(17): 5482–5515.*

La Rosa, M. Rizzo, R., Urso, A. & Gaglio, S. 2008. Comparison of Genomic Sequences Clustering Using Normalized Compression Distance and Evolutionary Distance. *Lecture Notes in Computer Science 5179: 740–746.*

Li, M., Badger, J.H., Chen, X., Kwong, S., Kearney, P. & Zhang, H. 2001. An information-based sequence distance and its application to whole mitochondrial genome phylogeny. *Bioinformatics 17: 149–154.*

Lin, J. 1991. Divergence measures based on the Shannon entropy. *IEEE T Inform Theory 37: 145–151.*

Liu, B., Lin, L., Wang, X., Dong, Q, & Wang, X. 2008. A Discriminative Method for Protein Remote Homology Detection Based on N-nary Profiles. *Bioinformatics Research and Development Second International Conference, BIRD 2008 Vienna, Austria, July 7–9, 2008 Proceedings. Eds. Elloumi, M. et al. Communications in Computer and Information Science 13(1(2)): 74–86.*

Liu, F., Tøstesen, E., Sundet, J.K., Jenssen, T-K., Bock, C. et al. 2007 The Human Genomic Melting Map. *PLoS Comput Biol 3(5): e93.*

Löytynoja, A. & Goldman, N. 2008. Phylogeny-aware gap placement prevents errors in sequence alignment and evolutionary analysis. *Science 320: 1632–1635.*

Lu, G., Zhang, S. & Fang, X. 2008. An improved string composition method for sequence comparison. *BMC Bioinformatics. 9(Suppl 6): S15.*

Luo, J. & Zhang, X. 2007. New Method for Constructing Phylogenetic Tree Based on 3D Graphical Representation. *Bioinformatics and Biomedical Engineering ICBBE 2007. The 1st International Conference, proceedings, pp. 318–321.*

Machado-Lima, A., del Portillo, A.H. & Durham, A.M. 2008. Computational methods in noncoding RNA research. *Journal of Mathematical Biology 56(1–2): 15–49.*

Macke, T. J., et al. 2001. RNAMotif, an RNA secondary structure definition and search algorithm. *Nucleic Acids Research 29(22): 4724–4735.*

Majewski, J. 2003. Dependence of mutational asymmetry on gene-expression levels in the human genome. Am J Hum Genet 73: 688–692.

Mani, P., Yadav, V.K., Das, S.K. & Chowdhury, S. 2009. Genome-Wide Analyses of Recombination Prone Regions Predict Role of DNA Structural Motif in Recombination. *PLoS ONE 4(2): e4399.*

Mantaci, S., et al. 2008. Distance measures for biological sequences: Some recent approaches. *Int J Approx Reason 47: 109–124.*

Martinez-Guerrero, C.E., Ciria, R., Abreu-Goodger, C., Moreno-Hagelsieb, G. & Merino, E. 2008. GeConT 2: gene context analysis for orthologous proteins, conserved domains and metabolic pathways. *Nucleic Acids Res. 36(Web Server issue): W176–180.*

Nikolaou, C. & Almirantis, Y. 2006. Deviations from Chargaff's second parity rule in organellar DNA Insights into the evolution of organellar genomes. *Gene 381: 34–41.*

Nix, D.A. & Eisen, M.B. 2005. GATA: a graphic alignment tool for comparative sequence analysis. *BMC Bioinformatics 6: 9.*

Obbard, D.J., Gordon, K.H.J., Buck A.H. & Jiggins, F.M. 2009. The evolution of RNAi as a defence against viruses and transposable elements. *Phil. Trans. R. Soc. B 364: 99–115.*

Pan, J., Deras, M.L. & Woodson, S.A. 2000. Fast folding of a ribozyme by stabilizing core interactions: evidence for multiple folding pathways in RNA. *J Mol Biol 296: 133–144.*

Parker, S.C.J., Margulies, E.H. & Tullius, T.D. 2008. The relationship between fine scale DNA structure, GC content, and functional elements in 1% of the human genome. *Genome Informatics 20: 199–211.*

Parker, S.C., Hansen, L., Abaan, H.O., Tullius, T.D., Margulies, E.H. 2009. Local DNA Topography Correlates with Functional Noncoding Regions of the Human Genome. *Science 324(5925): 389–392.*

Paten, B., Herrero, J., Beal, K. & Birney, E. 2009. Sequence progressive alignment, a framework for practical large-scale probabilistic consistency alignment. *Bioinformatics 25(3): 295–301.*

Perera, A., Vallverdu, M., Caminal, P. & Soria, J.M. 2008. DNA Binding Site Characterization by Means of Renyi Entropy Measures on Nucleotide Transitions *IEEE Transactions on NanoBioscience 7(2): 133–141.*

Pritchard, L., White, J.A., Birch, P.R.J & Toth, I.K. 2006. GenomeDiagram: a python package for the visualization of large-scale genomic data. *Bioinformatics 22(5): 616–617.*

Putonti, C., Luo, Y., Katili, C., Chumakov, S., Fox, G.E., Graur, D. & Fofanov, Y. 2006. A computational tool for the genomic identification of regions of unusual compositional properties and its utilization in the detection of horizontally transferred sequences. *Mol Biol Evol. 23(10): 1863–1868.*

Qi, J., Wang, B., &Hao, B. 2004. Whole proteome prokaryote phylogeny without sequence alignment: a k-string composition approach. *J. Mol. Evol. 58(1):* 1–11.

Russell, D.J., Otu, H.H. & Sayood, K. 2008. Grammar-based distance in progressive multiple sequence alignment. *BMC Bioinformatics 9: 306.*

Saha, S., Bridges, S., Magbanua, Z.V. & Peterson, D.G. 2008. Empirical comparison of ab initio repeat finding programs. *Nucleic Acids Res. 36(7):* 2284–2294.

Saunders, N.J. *et al.* 2000. Repeat-associated phase variable genes in the complete genome sequence of Neisseria meningitidis strain MC58. *Molecular Microbiology 37(1): 207–215.*

Schwartz, D., Chou, M.F. & Church, G.M. 2009. Predicting protein post-translational modifications using meta-analysis of proteome scale data sets. *Mol Cell Proteomics. (2): 365–379.*

Sims, G.E., Jun, S.R., Wu, G.A. & Kim, S.H. 2009. Alignment-free genome comparison with feature frequency profiles (FFP) and optimal resolutions. *Proc Natl Acad Sci USA. 106(8): 2677–26782.*

Sonnhammer, E.L.L. & Durbin, R. 1995. A dot-matrix program with dynamic threshold control suited for genomic DNA and protein sequence analysis. *Gene 167: GC1–10.*

Thirumalai, D. & Hyeon, C. 2009. Theory of RNA Folding: From Hairpins to Ribozymes In: *Non-Protein Coding RNAs. Springer. Chapter 2, pp. 27–47*

Tøstesen, E., Sandve, G.K., Liu, F. & Hovig, E. 2009. Segmentation of DNA sequences into two-state regions and melting fork regions. *J. Phys. Condens. Matter 21: 034109 (9pp).*

Tøstesen, E. 2008. A stitch in time: Efficient computation of genomic DNA melting bubbles. *Algorithms Mol Biol. 3: 10.*

Tóth, G., Deák, G., Barta, E. & Kiss, G.B. 2006. PLOTREP: a web tool for defragmentation and visual analysis of dispersed genomic repeats. *Nucleic Acids Res. 34(Web Server issue): W708–13.*

Touchon, M., Nicolay, S., Audit, B., Brodie of Brodie, E.B., d'Aubenton-Carafa, Y., Arneodo, A. & Thermes, C. 2005. Replication-associated strand asymmetries in mammalian genomes: Toward detection of replication origins. *PNAS 102: 9836–9841.*

Ulitsky, I. et al. 2006. The average common substring approach to phylogenomics reconstruction. *J Comp Biol 13: 336–350.*

Vinga, S. & Almeida, J. 2003. Alignment free sequence comparison - a review. *Bioinformatics 19: 513–523.*

Vinga, S. & Almeida, J.S: 2004. Rényi continuous entropy of DNA sequences. *J Theor. Biol. 231: 377–388.*

Vinga, S. & Almeida, J.S. 2007. Local Renyi entropic profiles of DNA sequences. *BMC Bioinformatics 8: 393.*

Vitányi, P.M.B., Balbach, F.J., Cilibrasi, R.L. & Li, M. 2009. Normalized Information Distance. In: *Theory and Statistical Learning, Springer, pp. 45–82.*

Wan, L., Li, D., Zhang, D., Liu, X., Fu, W.J., Zhu, L., Deng, M., Sun, F. & Qian, M. 2008. Conservation and implications of eukaryote transcriptional regulatory regions across multiple species. *BMC Genomics. 9(1): 623.*

Wang, A., Chang, G, & Hu, M. 2008. A Color Visualization Method to the Isochores of DNA Sequences. *Fourth International Conference on Natural Computation 5: 3–7.*

Xie, X., Rigor, P. & Baldi, P. 2009. MotifMap: a human genome-wide map of candidate regulatory motif sites. *Bioinformatics 25(2): 167–174.*

Xue, C., Li, F., He, T., Liu, G.P., Li, Y. & Zhang, X. 2005. Classification of real and pseudo microRNA precursors using local structure and sequence features and support vector machine. *BMC Bioinformatics 6: 310.*

Yang, J. & Deogun, J.S. 2008. Classifying G Protein-Coupled Receptors with Multiple Physicochemical Properties. *In: Proceedings of BMEI 2008, pp 93–97.*

Yau, S. S.-T., Wang, J., Niknejad, A., Lu, C., Jin, N. & Ho, Y.-K. 2003. DNA sequence representation without degeneracy. *Nucleic Acids Res. 31(12): 3078–3080.*

Yoon, K., Ko, D., Doderer, M., Livi, C.B. & Penalva, L.O. 2008. Over-represented sequences located on 3' UTRs are potentially involved in regulatory functions. *RNA biology 5(4): 255–262.*

Zhang, S.H. & Huang, Y-Z. 2008. Characteristics of oligonucleotide frequencies across genomes: Conservation versus variation, strand symmetry, and evolutionary implications. *Nature Precedings 2008.2146.1*

Zhang, Z.-J. 2009. DV-Curve: a novel intuitive tool for visualizing and analyzing DNA sequences. *Bioinformatics 25(9): 1112–1117.*

Zhaxybayeva, O. 2009 Detection and quantitative assessment of horizontal gene transfer. *Methods Mol Biol. 532: 195–213.*

Zielinski, J.S., Bouaynaya, N., Schonfeld, D. & O'Neill, W. 2008. Time-dependent ARMA modeling of genomic sequences. *BMC Bioinformatics 9(Suppl 9):S14.*

Chapter 5

Molecule Structure Based Searching and Comparison

Massive amounts of information keep accumulating into various chemical structure databases about complex molecules, including protein and RNA structures, drug molecules, drug-ligand databases, and so on. Surprisingly, there is no commonly accepted standard for recording and managing chemical structure data, *e.g.* drug molecules, which would be suitable for automated data mining (Banville 2006). However, the data formats used in the largest public database, PubChem, seem to be becoming a *de facto* standard. PubChem contains over 40 million entries, and can be searched by text queries or structural similarity queries using SMILES formulas and other similar text string based structure representations (Hur & Wild 2008). There is also a multitude of small molecule databases for drugs and metabolites available on the internet, which already have over ten million entries with 2D image representations of structures (see the extensive database review of Jónsdóttir *et al.* 2005).

Also structural data on DNAs and especially on RNAs folded into 2D and 3D structures is accumulating at increasing speed. For example, tens of thousands of 2D folded non-coding RNA structure candidates along the human genome have been computed and are available on the internet in the miRNAMap database (Hsu *et al.* 2006). For proteins, there are about 50,000 3D detailed structures (both crystallography and magnetic resonance based atomic resolution data) in the Protein Data Bank PDB (http://www.wwpdb.org/). A multitude of structures have also been computed by *ab* initio folding programs or by sequence threading to a known protein structure, and many websites exist for comparing such structures by a variety of algorithms. These are outside the scope of this book and too numerous to review here.

In the simplest traditional approach for finding proteins with similar structures, amino acid sequence similarity can be used as a proxy to similarity. However, a minimum of 30% sequence identity and a known structure is needed for reliable finding of remotely related protein structures, because protein structure is known to be much more conserved than sequence similarity. For accurate drug design, up to 60% sequence identity is needed to ensure proper ligand binding models in molecular simulations. Furthermore, the known set of protein structures does not appear yet to cover sufficiently the natural protein structure space (Vitkup *et al.* 2001), so that pre-computed structural indexes will not cover all the novel proteins that could be discovered or synthesized.

A recent analysis of 203 genomes (including 17 eukaryotes) indicates that merely to cover the 2,000 or so structural domain families in CATH database, some 1,000 well selected structures need to be determined. However, to get accurate models for all natural proteins, about 90,000 solved structures are needed for good modelling starting points of more than 30% sequence similarity (Marsden *et al.* 2006). We still lack sufficient data, although the largest current database of modelled structures MODBASE (based on related known structure proteins) contains already over 5 million computed domain models for 1.5 million different proteins (Pieper *et al.* 2009). Note that both modelling and searching this database is based on multiple sequence alignments of protein sequences known to be related. Thus fast and accurate *de novo* structure based searches for *e.g.* protein analogs is still needed.

The high-accuracy sophisticated structural alignment algorithms are crucial for drug design, *e.g.* ligand to protein binding simulation. However, these algorithms currently cannot handle simultaneous comparison and classification of large numbers of structures, except by brute force using very large distributed computing infrastructures, such as Proteins@Home and Rosetta@Home hosted on the BOINC grid computing facility (boinc.berkeley.edu). Thus there is currently no efficient solution for matching, clustering and classifying post-genomic scale numbers of protein molecular structures.

Nowadays three main methods of protein sequence-based structural alignment are used. They all rely on pairwise similarity measures and the most important are the algorithms VAST, SHEBA and DALI (for algorithmic details, see Sam *et al.* 2009). However, they are all quite slow to compute and still do not accurately reflect the gold standard of manually created protein domains in SCOP or CATH protein domain databases. The NCBI database of pre-computed protein neighbours uses the VAST pairwise algorithm, which has been recently speeded up markedly (up to 100-fold) by Malod-Dognin *et al.* (2009) by clever search optimization for multiple CPU processing.

However, multiple structural alignments still take very much longer than pair-wise alignments, even with the fastest recent method (Margraf & Torda 2008), which took 8 hours on one CPU to align structures of only 818 proteins in a large protein family. The recent protein segment finder of Samson & Levitt (2009) also needs extensive pre-computations to achieve a quickly accessible search index based on known proteins with known structures.

For smaller datasets, successful prediction of desired structure-derived functionalities from 3D structural data appears quite feasible, using a variety of well known methods, which are important for the pharmaceutical industry studying small families of related drug candidates or proteins. For example, Kohonen SOMs were used to predict cytotoxicity with 89% accuracy for 55 terpenoid compounds based on geometrical 3D features (Fernandes *et al.* 2008). However, in general, fast 3D structure comparisons are not yet speedy enough for post-genomic large-scale known protein universe comparisons. One approach to tackle this problem is to subdivide the problem in some way. As an example, recently Aung & Tan 2005 used hierarchical filtering and elimination of unlikely groups of structures based on pre-computed diagnostic structural features, but, again, this demands prior knowledge about the 3D structures of the molecular structures.

From the foregoing it is clear that, apart from small-scale case studies with limited datasets, current 2D and 3D protein structure comparison methods are not fast enough for comparing on-the-fly the millions of various chemical structures already available and expanding at increasing pace. There is still a great need for alternative faster approaches. The following subchapters explore some of the more interesting recent novel methods. A comprehensive treatment of improving the conventional 3D matching methods is outside the scope of this book.

In subchapter 5.1.1 methods of coding 3D structure to 1D strings is introduced and in subchapter 5.1.2 methods using graph-like features from 3D or 2D structures are reviewed. Subchapters 5.1.3 and 5.1.4 introduce various graph visualisation methods and graph grammars, respectively. Finally, in subchapter 5.2 new approaches based on in silico extracted structural features are exemplified using RNA structure comparisons.

5.1 Molecule structures as graphs or strings

5.1.1 *3D to 1D transformations*

One way to get around the computational bottleneck of comparing complicated 3D structures is to use structure-related alphabets to code 3D structural information into a linear string of symbols and then to search for similar encoded strings by a faster text-based search. Le *et al.* (2009) used an alphabet of 20 symbols for 20 different short structural motifs, each 4 amino acids long, to code 3D structure to a 1D character string. Surely enough, comparing the structure-encoded strings is much faster, but less accurate than true 3D structure alignment. These methods appear more accurate for structural similarity than the protein coding amino acid sequence similarity searches by *e.g.* traditional BLAST. However, the 3D to 1D transformative methods still need significant improvement to be competitive in their predictive ability with true 3D alignments (Le *et al.* 2009).

One interesting new approach of recoding 3D information to 1D is Ramachandran codes method, which has been stated to approach BLAST similarity search speeds for conserved protein structures (Lo *et al.* 2007), while having better precision and recall than BLAST. Briefly summarized, all protein backbone structures are characterised by three torsion angles (Figure 5-1, top), two of which, phi and psi, can be used to generate a 2D plot for the angular values at successive amino acids (Figure 5-1, bottom). The plot summarises the 3D structure of a protein in 2D space. This 2D image can be further compressed into a 1D string called a Ramachandran code by assigning a letter to various areas in the 2D plot which contain clusters of angle pair points. The linear codes for different proteins can then be compared using standard string comparison methods, like BLAST *etc.* The resulting alignments of Ramachandran code strings show more clearly the structural similarities than amino acid sequence alignments (Figure 5-2).

Ramachandran codes is an example of a promising 3D to 1D transformation of protein structure data. This coding method has been utilised to obtain compressed data for proteome-scale fast detection of circularly permuted protein variants (Lo & Lyu 2008). This approach is interestingly reminiscent of the Burrows-Wheeler algorithm discussed in a later chapter 9.8 on GPU acceleration of database searches. Obviously, a prerequisite of using the Ramachandran codes method is the availability of the initial 3D structure data for the torsion angles phi and psi.

Another approach to coding 2D information is to make a string of 0's and 1's based on the presence or absence of a 2D feature from a list of basic informative structural units, like a ring of 6

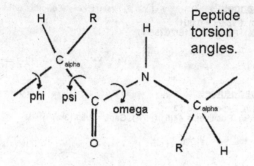

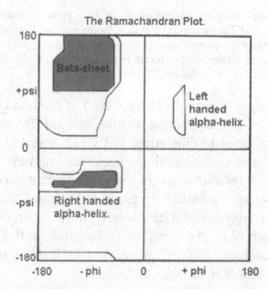

Figure 5-1. 2D Ramachandran plots based on torsion angles phi and psi (top) show clustering of paired angle values based on 3D structures (bottom). Reproduced from http://swissmodel.expasy.org/course/text/chapter1.htm.

carbons, a linear segment of 3 carbons with double bonds and so on. The PubChem Fingerprints are based on this idea, and the string currently contains 881 bits for 881 different kinds of basic units occurring in the deposited molecules. The similarity of the fingerprints is then a good indication of the structural similarity.

Li (2006) used this type of coding, but with frequencies of the basic units included, to cluster millions of compounds in a few

Amino acid sequences: d1b3aa_ (Length = 67) vs. d1tvxa_ (Length = 64)

Score = 21.3 bits (58), Expect = 0.001
Identities = 5/29 (17%), Positives = 16/29 (55%)

```
d1b3aa_:   2 PYSSDTTPCCFAYIARPLPRAHIKEYFYTSGKCSNPAVVFVTRKNRQVCANPEKKWVREYINSLEMS- 68
                                      C+   V+   +   R++C +P+    +++ +
d1tvxa_:  23 ----LRCLCIKTTSGIHPKNIQSLEVIGKGTHCNQVEVIATLKDGRKICLDPDAPRIKKIVQKKLAGD 86
```

Ramachandran sequences: d1b3aa_ (Length = 65) vs. d1tvxa_ (Length = 62)

Score = 52.0 bits (119), Expect = 7e-13
Identities = 18/51 (35%), Positives = 44/51 (86%), Gaps = 2/51 (3%)

```
d1b3aa_:   3 QFNFFILIHNIGFBILFLCKKHNIIIH--FFKEFDGFSHHHHQKNPFMGLILDKFCBABAACCABKN- 67
                                FL++K++II+HI  F  +  ++++HHHQ+N+F++L+LDK++CA+BB+BB+  +
d1tvxa_:  24 ------QFELNGIINNFLDDKEEIIGHGQFICNMNILGMHHHQDNWFGILGLDKLDCADBBABBABKP 85
```

Figure 5-2. Comparison of an alignment of two proteins using amino acid sequence (top) and Ramachandran codes (bottom); the Ramachandran code letters show clearly a stronger similarity (longer matching segments) than the amino acid coding letters. Reproduced from Lo *et al.* (2009). Reproduced by Open Access policy of the publisher BioMed Central.

hours on a single standard PC. This kind of methods do have the requirement of first extracting all the relevant 2D features from the molecules prior to clustering, but given a stored database of previously examined compounds, accumulating new information for new chemicals and searching similar structures can be made fast and efficient. Recently, Trepalin & Yarkov (2008) clustered such subunit frequency data hierarchically to enable fast and accurate search of all the 4 million compounds of the Aurora fine chemicals database. Finally, chemical text mining methods are reviewed by Sun *et al.* (2008).

5.1.2 *Graph matching methods*

Matching graphs is an old problem in topology and network theory, including analysis of internet and social networks. Many aspects of established graph analysis theory have been already applied to biodata mining and analysis. Only illustrative examples of this wide field are included here, but graph kernels are discussed further in chapter 6.1 on SVM kernels. Good general reviews of various other graph matching methods in this very active field are in Washio & Motoda (2003) and Han *et al.* (2007).

Molecule 3D structures can be represented in 2D in two basic ways: as graphs or networks of elements (discussed in this subchapter), or bitmap images both for visual consumption for the human eye and for processing by image analysis algorithms (discussed in the subchapter 5.3). When molecular structures are represented as graphs of connectivity between elements of the molecule, those graphs can be compared and clustered using various graph matching methods, visualized in a variety of ways or described in terms of graph grammars, examples of which are discussed below. In these methods the information about the graph structure has to be transformed to suit the clustering or classification algorithm.

Faulon (2004) has developed a powerful method to encode molecular structure connectivity to a so-called molecular signature. Graphs of molecules in 2D indicate the proximities and paths (linkages) between the atoms or subunits of the molecule. Each atom is surrounded by a **subgraph** of neighbouring atoms. The atomic signature can be defined as a subgraph including all atoms and bonds up to a predefined distance from the given atom. This data is represented by a molecular signature, which is a vector of occurrence numbers of all the atomic signatures in a molecule. The stepwise vector space then encodes the 2D structure of the whole molecule by the combination of subgraphs of the individual atoms.

The vectorized data for each chemical can then be used as input to *e.g.* SVM classification to compare in genome-scale large numbers of protein structures. (Faulon *et al.* 2008) developed the method further by creating signature kernels (kernel methods are discussed further later on in this book in conjunction with SVM methods, see chapter 6.1) which can classify fast thousands of chemicals and even simultaneously utilize linked information about known chemical reactions produced by the chemicals.

The signature kernel method essentially replaces traditional protein homology comparison of amino acid alignments with atomic level data encoded to signature vectors. Importantly, the signature vectors makes more information available to clustering and classifying than mere protein homology data. The signature kernel method thus appears powerful for prediction of reaction-enzyme and drug-target interactions and compares well with other state of art methods like SMILES string comparisons (see below) and even sophisticated kernels based on 3D atomic distances (Swamidass *et al.* 2005).

The great advantage of the signature product kernel is that it can predict cases even when neither the protein sequence nor the chemical reaction are available for the training dataset, and still reach an accuracy of about 80%. The method was recently used successfully for the prediction of inhibitors of coagulation factor Xia, and the developed SVM classifier could quickly produce predictions for all the 12 million compounds in PubChem (Weis *et al.* 2008). Recently, the complexity of the molecular signature was increased by using real numbers instead of integers to code the vector space of connections, in order to encode more information about the subgraphs into the signature vector (Catana 2009). This improved classification in some test cases, but further development and testing is necessary to show whether the added complexity in the algorithm actually increases classification performance for all kinds of datasets.

In more recent work along the same lines, Jacob *et al.* (2008) developed various 2D structural graph based kernels based on 8-step long subgraphs, which predicted well ligands and drug-target interactions for even novel molecules. The best performing kernel included hierarchical information about the enzyme EC numbers with enhanced prediction accuracy of over 90%. Jacob *et al.* term their approach chemogenomics and showed that the method performs very well for even the GPCRs (G-protein coupled receptors) which are considered a difficult family of proteins to predict functions for (Jacob & Vert 2008). The SVM graph kernel approach might even be used to predict quantitatively the affinities and reaction kinetics of the predicted drug-target or enzyme-substrate pairs. For more information, see the good review of SVM related protein and small molecule kernels by Vert & Jacob (2008).

As for the earlier traditional methods for encoding molecular structures to strings, the SMILES notation (reviewed in Yadav *et al.* 2004, and used *e.g.* by PubChem) is an established transformation from 2D graph structure to 1D string. Several variants of the notation language have been developed and also commercialised by the Daylight company (www.daylight.com). It is well suited to exact graph matching searches, but may not be so suitable for approximate match searches or clustering which depends on quantitative differences between the molecular structures. Also, fast query performance comes at the cost of large pre-processing computation.

GraphGrep is another recent graph approximate match finding algorithm, introduced by Giugno & Shasha (2002). It is fast and efficient, but does not allow customized fuzziness in specified locations in the molecule, nor does SAGA (Tian *et al.* 2007), another modern and flexible graph approximate matching system. In an enhancement to GraphGrep, GraphFind (Ferro *et al.* 2008) allows one to specify which parts of the query molecule are required to match exactly with target graphs and which parts are allowed to vary from the query. Other new approaches are also emerging, notable among them the G-hash, which addresses the problem of fast indexing of the matching graph candidates by a hashing method linked to a novel graph kernel to enable fast queries and access to pre-computed similarities (Wang *et al.* 2009).

However, as a final note to this subchapter of graph matching, often the detailed graph data is not available for all the molecules in the dataset. Also, finding all the exact or even approximate matches to (sub)graphs of interest may not be necessary, if other data is available. For example, Bruckner *et al.* (2008) found that combined sequence similarity and clustering of the set of target proteins in a network of protein interactions within a specified number of steps was enough to identify metabolically and regulationally related groups of proteins between yeast and human. This was achieved without the slow approximate search for matching subgraphs. Therefore it is always good to ask 1) whether the data at hand is sufficient (*i.e.*, are enough subgraphs available in the database) and 2) whether the tedious subgraph quantitative matching is really necessary, if other kind of data is available.

5.1.3 *Graph visualization*

While a multitude of academic freeware has been developed for specific network visualization purposes, no single system has been widely adopted as a standard method. The commercial package Cytoscape (Cline *et al.* 2007) is recently gaining popularity in systems biology research and appears to be quite flexible to accept various kinds of inputs ranging from network node pair lists to XML and RDF triplet type data.

Academic open source software systems for various specific applications include the handy Java-based InterViewer (Han & Ju 2003, Han *et al.* 2004). For recent reviews of such tools, such as CytoScape, BioLayoutExpress, Medusa, PATIKA, PAJEK *etc.*, see Suderman & Hallett (2007) and Pavlopoulos *et al.* (2008). The general impression is that these systems still suffer from lack of standardization, parallel development of overlapping and heterogeneous functionality and variable quality of visualization for networks, labelling and annotation as well as inflexible import/export format limitations. One generic problem is also the lack of commonly adopted standards for encoding complex annotated network data.

A new advanced and more integrated data storage and visualization package for network biodata is the ViSant system (Hu *et al.* 2008). An example snapshot from ViSant is shown in Figure 5-3. In a multi-center collaboration with KEGG database, metabolic pathways have been included in the system, and the Predictome database and visualization system incorporates known molecular interactions from various interaction databases like MINT, BIND, MIPS as well as computed predictions of the interactions. Predictome currently contains some 600,000 interactions from over 100 different organisms, half of which are

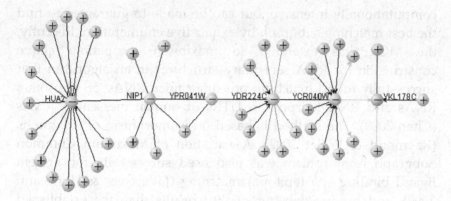

Figure 5-3. A series of expanded nodes in yeast gene interaction network visualized by Visant. From the Visant manual, at http://visant.bu.edu/vmanual/.

experimental, half computer-predicted. For example, in the section for the human metabolism, some 87,000 interactions are included.

5.1.4 *Graph grammars*

While molecular structure fingerprints are fast, they cannot show the location of the matched graph on the found target, *i.e.* the matching subgraph. More direct methods are needed for this and graph grammars are one of such methods. Graph grammars refer to lexical notation systems for representing graphs in serially constructed symbol strings. Only recently have graph grammars gained popularity in the description of biomolecules and metabolic pathways in systems biology, as reviewed in Rossello & Valiente (2005). These methods are likely to become important in large-scale mining of complex biomolecular interaction networks. Alternatives to structure algebras are SVM based graph kernels, discussed later in chapter 7.1.

In the general area of graph grammars, the traditional approach is to match subnetwork motifs within one network or between two or more networks by Maximum Common Subgraph. These Maximum Common Subgraph (MCS) methods are computationally intensive, but can be made to guarantee to find the best matching subgraph by exhaustive enumeration. Recently, the MCS was extended to maximum compatible clique construction for RNA secondary structures in an algorithm that successfully found matching precursor microRNAs, 5S ribosomal RNAs and RNA hairpins in UTRs of human messenger RNAs (Chao 2008). The method is based on isomorphism of subgraphs (Raymond & Willet 2002). A variation of Maximum Common Subgraph isomorphism was also used successfully for protein ligand binding site (epitope) matching (Jakuschev & Hoffmann 2009), and produced better quality results than the established standard algorithm ASSAM (Artymiuk *et al.* 2005).

The MCS method is still under active development and recently a more computationally efficient back-tracking version of the algorithm was developed for predicting functions of drug-like molecules (Cao *et al.* 2008). Their algorithm output was useful as input to SVM in conjunction with other structural features and outperformed other current state-of-the-art systems. Such methods may be useful for RNA structures as well, discussed in more detail in the subchapter 5.2 below.

Recently a new alternative to MCS has emerged: Graph Isomorphism, which is the mapping of graph nodes (vertices) in two complete graphs (not subgraphs, like in the method of Chao mentioned above) to each other to maximize the number of matched nodes. If all nodes can be matched, the two graphs are said to be completely isomorphic. Graph isomorphism was recently used for analyzing metabolic pathways (Crabtree & Mehta 2009). This method may become a competitive alternative to MCS methods in biological network analysis, as practical implementations start to appear.

5.2 RNA structure comparison and prediction

The recently discovered microRNAs have given added impetus to the analysis and derivation of 2D and 3D structures of folded

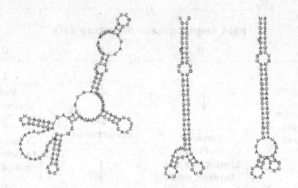

Figure 5-4. Typical microRNA 2D structures (middle and right) and a random genome segment structure (left), folded by RNAfold of Vienna package.

RNAs. The 2D and 3D structures of various ribosomal and other small RNAs have already been studied a long time, though often without the modern rigor of graph theory. *In silico* heuristic folding of such structures is already routine for at least smaller molecules and the Vienna package is the standard tool for this. Examples of *in silico* folded microRNA structures obtained by RNAfold are shown in Figure 5-4.

The basic approaches to comparing RNA structures are depicted in Figure 5-5. Starting from a set of RNA sequences, the traditional approach is to align them to get a consensus RNA sequence (path A in Figure 5-5), which then can be folded *in silico* using standard methods, like RNAfold from the Vienna package developed by Ivo Hofacker (http://www.tbi.univie.ac.at/~ivo/RNA/) or Mfold of Zuker (server at Rensselaer Polytechnic Institute at http://www.bioinfo.rpi.edu/applications/mfold/).

For a good recent review on the large number of available 2D structural *in silico* folding algorithms and the smaller number of new 3D algorithms, see Capriotti & Marti-Renom (2008). In the implementation of path A in Figure 5-5 most standard DNA alignment methods like FASTA, BLAST, T-COFFEE *etc.* can be used in a routine manner. The BLAST scoring matrices can be

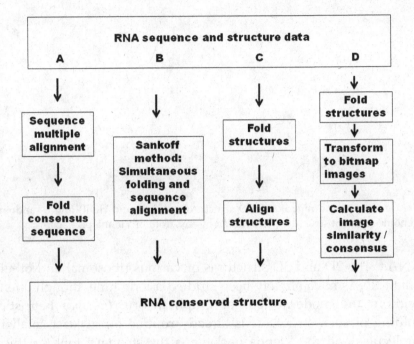

Figure 5-5. Different approaches to RNA conserved structure analysis.

modified to obtain better accuracy for evolutionary structural similarities between RNAs.

Method A suffers from the problem of not clustering together all related sequences, as RNA structure is more conserved than its sequence, as is the case for proteins.

Method B is the other classical approach, based on simultaneous multiple alignment of sequences and 2D structures and is called the Sankoff method (Sankoff 1985). However this method does not allow clustering of sequences/structures and has prohibitive time complexity of computation for longer sequences and larger numbers of sequences to be aligned.

This computational bottleneck in the Sankoff method has in fact driven the development of the other alternatives in Figure 5-5, though no clear winner algorithm appears to have emerged yet. The Sankoff algorithm may actually be in for a revival, because a

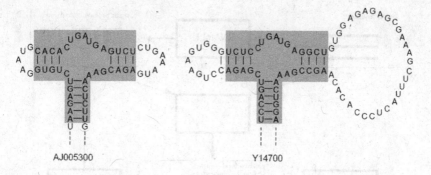

Figure 5-6. A local substructure of folded 2D RNA structure obtained by LocaRNATE. Reproduced from Otto *et al.* (2008) by permission from IEEE, © 2008 IEEE.

speeded-up version called LocaRNATE (Will *et al.* 2007) has been developed recently. This method could be usefully implemented for GPU processing (for more on GPU computing, see chapter 9.8). The LocaRNATE approach has been utilized also for local matching of substructures (Otto *et al.* 2008, for its workflow, see Figure 5-7), which can aid in building the correct multiple alignment of both sequences and structures.

The third alternative (path C in Figure 5-5) is to rely on folding candidate 2D structures first and then aligning structures by various methods, including *e.g.* the subgraph alignment methods discussed above. Path C has been explored by many research groups, and several microRNA structure databases are based on this approach. Algorithms for method C depend crucially on the method to align structures, and their improvement is still going on. Many algorithms are based on the concept of RNA as topological graphs (Gan *et al.* 2004) or trees. Among notable early algorithms are RNAFORESTER (Höchsmann *et al.* 2004), MARNA (Siebert & Backofen 2005), TREEMINER (Zaki *et al.* 2005) and RSMATCH (Liu *et al.* 2005). Their performance in analyzing and clustering very large RNA sets has not yet been compared.

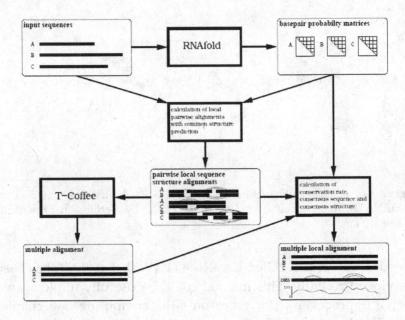

Figure 5-7. The flowchart of LOCARNATE algorithm. Reproduced from Otto *et al.* (2008) by permission from IEEE, © 2008 IEEE.

Stem loop structure analysis is useful also for genomic DNA data. Cozzuto *et al.* (2008) analysed 40 bacterial genomes for stem-loops. Two thirds of the detected stem loops have a consistent conserved structure and are located in the intergenic regions, presumably involved in genome function regulation and as RNA genes not coding for proteins. Such stem loops are associated with various repetitive domains, as also found in De Gregorio *et al.* (2009), who analyzed two species of *Enterococcus* bacteria for palindromic repeats. The conserved short stem loops were mostly located near stop codons of genes, and appear to function as transcription terminators. Thus structural analysis of the genomes is really starting in earnest and many new discoveries are expected to be made for novel types of functional elements in both bacterial and eukaryotic genomes.

5.3 Image comparison based methods

A quite different approach for molecular structure comparisons is that of path D in Figure 5.5 which involves moving the problem from string analysis to image analysis domain. Two-dimensional images of molecular structures can be transformed into bitmap images and then analyzed by various image clustering and feature extraction methods (Havukkala *et al.* 2006). Two examples for microRNA analysis are discussed below in the later subchapters, but this kind of approach is generic for many kinds of biomolecular structures. This generic approach appears attractive especially for large scale processing of millions of pictures, as the technology of fast image comparisons for internet image searches is maturing. A relevant related method is the semantic hashing method for image data, discussed in subchapter 9.8.

In principle any image processing algorithm will do, as long as it captures the biologically relevant features from the 2D image of the represented molecule. For a review of the large variety of available shape feature extraction and comparison methods for images, see Zhang & Lu (2004) and Mingqiang *et al.* (2008). Yankov *et al.* (2008) reports an interesting new rotation invariant method of converting shapes into time series, one of the many attractive directions in this very active research field.

The general idea of transforming a shape comparison problem into a bitmap image comparison has the advantage that current advanced image analysis methods can extract a large number of possible features from the images, even those not recognized by the researcher initially. Such extracted features can then be combined with any other easily computed features from the molecular structure, like surface area or perimeter of the molecule, hydrophobicity *etc.* All these features can then be input to the desired classifier, including SVM.

In this way the structural clustering and classification problem solving can utilize the full range of artificial intelligence and data mining methods developed for large, high-dimensional datasets. The additional benefit is that image analysis may discover some features that are not obvious to the researcher or are not yet known based on traditional chemical or structural features normally extracted from the molecular structures.

One traditional image analysis method is rotation-invariant description of images using Gabor filters. They are based on rotational sampling of the image contrasting edges or other relevant features (for a recent review of Gabor filter methods, see Kamarainen *et al.* 2006). Gabor filters have been found to be most useful in complex image analysis.

A multitude of other types of shape analysis methods has been developed. Notable among them are shock graphs, which compare silhouette shapes by the amount of deformation of one shape to match another shape (Sebastian *et al.* 2004). This method, however, does not appear to have been tried for bioinformatics applications. Another popular method among image analysts is curvature scale spaces, which are based on crossings of the actual silhouette across the mean curve along the shape (Mokhtarian & Bober 2003). Curvature scale spaces have been used in an interesting application of classifying diatomic algae shapes (Jalba *et al.* 2005). Similar methods have been used to find unusual shapes from large databases in linear time (Wei *et al.* 2008). These methods appear well worth exploring for more extensive use in bioinformatics applications as well.

As for 3D image analysis, this field is relatively new in bioinformatics. However, in the engineering sciences many different tools have already been developed, both for surface representation (*e.g.* VRML) and CAD-based voxel representations. For a recent review, see Tangelder & Veltkamp (2008). These methods have not yet percolated into bioinformatics to make an impact, apart from proprietary commercial drug design and enzyme/receptor ligand binding modelling for therapeutics.

These new 3D methods are bound to become more widely used, as soon as more extensive 3D data at different resolutions from cells and biological tissue samples becomes available. This new kind of data includes chromosomal 3D structures and internal membrane structures within a cell during cell growth and during the cell division cycle. An interesting example in this direction is the angle-distance method of Chu *et al.* (2008) for the analysis of 3D images of proteins.

Finally, the reverse approach of moving from linear sequence data to 2D images should be mentioned. Long linear sequences of DNA or RNAs can be represented as walks in 2D or 3D space, as discussed in chapter 4.2.3 earlier, and then displayed as a 2D bitmap image in a fractal space. In this way large amounts of information can be compressed to a very compact 2D visual representation, which could then be analysed by bitmap image analysis methods. This approach for biodata displayed in fractal space does not appear to have been tried yet.

For biomedical applications, many sophisticated cross-platform image analysis software systems are already available. For example, ImageJ (Abramoff *et al.* 2004) is a Java-based biomedical image processing software tool that can interconvert a large number of various medical and microscopy image formats and can perform many image partitioning, segmentation and object identification tasks.

The image-processing toolbox of MatLab is also powerful, as well as the ITK and DipImage toolboxes. A good review of bioimage software tools and biomedical image databases is available (Peng 2008). Similarly advanced organism identification tools are also already available for *e.g.* insect identification (Mayo & Watson 2006, Cho *et al.* 2007 and O'Neill 2007). As interesting as these fields are, space in this book does not allow further delving into these algorithms (but see the related discussion on semantic hashing in chapter 9, as applied to large-scale image identification in general).

Before we go on to the two examples of bitmap image derived similarity assessment methods using RNA structure comparisons, it is interesting to note that an opposite approach is also possible, One could use the **dissimilarity** between objects to be classified, focusing on the distinguishing features, rather than first clustering similar items together for a training set for the classifier. The latest paper on this approach looks promising (Kim & Gao 2008), although no bioinformatics data appears to have been analyzed by this method. The theoretical aspects and application examples of this vectorization of graph information to a dissimilarity matrix are discussed in Bunke & Riesen (2008) and Riesen & Bunke (2009). Such methods could be utilized in biomolecular structure classification as well and await exploration.

5.3.1 *Gabor filter based methods*

Gabor filters for 2D images extract line segments of different orientations from the image based on rotational sampling of the picture at different angles. They have been used successfully for a large variety of image classification tasks, including fingerprints, faces and so on, although the method is considered to be computationally quite costly, precluding really large-scale applications (but see a recent development in speeding up the algorithm in Ilonen *et al.* 2007).

In essence, Gabor wavelets extract different information based on the angles of contrasting lines in the image, as shown in the example of Figure 5-8 for one microRNA in a 256 x 256 pixel image. The spatial frequencies at specified frequency and rotation angles result in a 2D array of magnitudes of signal in each image. The feature vector for classification is an ordered listing of these magnitudes for the image, and images are compared by summed

Figure 5-8. The folded RNA structure (left) processed by Gabor filter at 0, 90, 180 and 270 degrees directions (left to right). The direction dependent features are clearly extracted from the images. Adapted from Havukkala *et al.* (2005). Reproduced by permission from American Scientific Publishers.

differences of each element in the ordered feature vector. An all versus all matrix of similarities can then be calculated for an exhaustive comparison of all images.

On a sample set of 222 known microRNAs, the Gabor filter identified many correct pairs of structures known to be related by sequence similarity (Havukkala *et al.* 2005). In addition, many structural similarities were found that do not correspond to known microRNA families. Further tuning and testing of the method is, however, needed to evaluate the biological informativeness of the Gabor filter based clustering. One caveat is that the images are assumed to be aligned to the same orientation, and not *e.g.* mirror images of each other. Also, the speed of computation needs to be improved for the method to be scalable to larger datasets. It appears likely that alternative methods will be more efficacious (see *e.g.* the semantic hashing in chapter 9).

5.3.2 *Image symmetry set based methods*

Image feature extraction methods are still a rapidly evolving field, despite having been an active research area for decades. Some of the current methods have already been mentioned in chapter 5.2 above. A recent new approach for 2D shape similarities is based on symmetry sets (Kuijper *et al.* 2006) and other image symmetry detection systems are emerging from recent research (see *e.g.* Berner *et al.* 2008), which could be used for complex 2D or 3D biomolecule image analysis.

In the 2D image symmetry approach, the 2D shape is represented by a generalization of the Medial Axis method, where mutually symmetric points on the shape curve are connected by circles of varying size to generate a 2D array of values, which can be easily compared. One microRNA and its corresponding symmetry set characteristic diagram are shown in Figure 5-9. The 2D arrays can then be multiplied pointwise to achieve a summed similarity value between the two images. The method could identify similar shape microRNAs from a test set of 100

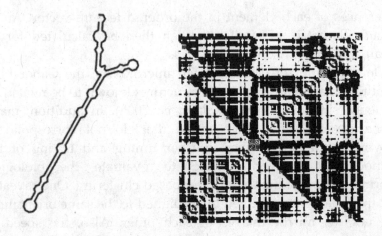

Figure 5-9. One microRNA and its 2D characteristic diagram derived from the image symmetry set. Adapted from Kuijper & Havukkala (2007). Reproduced by permission from IASTED/Acta Press.

microRNAs, regardless of their orientation, shown in Figure 5-10. Similar shapes on microRNAs were retrievable, including mirror image forms of the shapes. Rotational invariance is a very desirable feature in any image recognition method. This approach and similar methods would be worth exploring further with larger datasets and with optimized computation algorithms to speed up performance. The attraction of the method is that it involves as its main step matrix multiplication, which is easily parallelized and could suit the modern GPU computing platform being developed (see chapter 9.8).

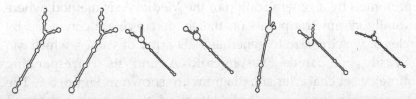

Figure 5-10. Examples of folded microRNA shapes, the query sequence is leftmost, followed by the best ranked candidates by the image symmetry method. Note that rotated similar shapes are also found. Adapted from Kuijper & Havukkala (2007). Reproduced by permission from IASTED/Acta Press.

5.3.3 *Other graph topology based methods*

RNA structure folding *in silico*, compact informative representation of the RNA structures and their clustering and classification are all very active research fields. Therefore only a few examples can be given here of the variety of approaches being taken.

In the area of *in silico* folding, Boltzmann ensembles of the RNA structures are sets of alternative structures in which the molecule spends most of its time. To decide the correct conformation or the most likely ones out of several alternatives is not easy. RNAfold has been the standard method of folding RNAs, but only outputs the best conformations ranked in binding energy order. Recently improved procedures for the alternative structures and their likelihoods have been developed, based on centroid estimation of high-dimensionality datasets (Carvalho & Lawrence 2008). The latest RNA folding method CONTRAfold utilizes such centroid estimators (Hamada *et al.* 2009) and improves the accuracy of predicted RNA structures substantially. This method is likely to be widely adopted, provided that the computational complexity can be obviated by further efficiency gains in running the algorithm.

Another relatively recent trend is to regard RNA structures as 2D graphs (Gan *et al.* 2004) or trees, as already discussed for RNA structural comparisons to guide the folding process. The graph can then be compared with a variety of methods, depending on the representation method of the structural information of the folded RNA. Typically the structural information is obtained from the 2D picture or bracket equation outputted by RNAfold.

As this research area is very active, new interesting structure representation methods keep appearing (for theory, see Knisley & Knisley 2008). One of them is XIOS RNA graphs (Li *et al.* 2008), which show RNA structures as networks of vertexes, in which each base-paired stem is a vertex connected to other stems. The connections are of different types, either exclusive, inclusive, overlapping or serial, depending on the topology of the binding segments within the folded structure. A depth-first-search is then done on the target graph. Using a dictionary of pre-calculated structures, a very fast structure search can be performed.

Most recently, the much earlier developed topological indexing methods from computational chemistry (Balaban & Ivanciuc 1999) are finally being adopted in molecular biology, as evidenced by the new graph data formulation to represent RNA structures by element contact graphs (Shu *et al.* 2008). In this work three topological indexes based on element contact graphs (Wiener index, Balaban index and Randić index) were tested and were found to be very useful in predicting the class of true non-coding RNAs based on structure, with an accuracy of 96–98%.

As a final note to the RNA 2D structure comparisons, there is no shortage of already computed structures, *e.g.* MirOrtho database contains already thousands of structures for good microRNA candidates, which await large-scale structural clustering and comparison. The onus is now on the bioinformaticians to choose the best algorithms to make sense of these large repositories of candidate RNA structures to screen out the best drug leads and to discover the functional structural features maintained by evolution in different organisms.

References

Abramoff, M.D. *et al.* 2004. Image processing with ImageJ. *Biophoto. Int. 11: 36–42.*

Artymiuk, P.J., Spriggs, R.V. and Willett, P. 2005. Graph theoretic methods for the analysis of structural relationships in biological macromolecules. *Journal of the American Society for Information Science and Technology, 56(5): 518–528.*

Aung, Z. & Tan, K.-L. 2005. Automatic 3D Protein Structure Classification without Structural Alignment. *Journal of Computational Biology. 12(9): 1221–1241.*

Balaban, A.T. & Ivanciuc, O. 1999. Historical Development of Topological Indices. *pp. 21–57 In: Devillers J and Balaban AT. (eds) Topological Indices and Related Descriptors in QSAR and QSPR. Netherlands, Gordon and Breach Science Publishers.*

Banville, D.L. 2006. Mining the chemical structural information from the drug literature. *Drug Discovery Today 11(1/2): 35–42.*

Berner, A., Bokeloh, M., Wand, M., Schilling A. & Seidel, H.-P. 2008. A graph-based approach to symmetry detection. *In: Proc. Symp. On Volume and Point-Based Graphics 2008, H.-C. Hege, D. Laidlaw, R. Pajarola & O. Staadt (Editors).*

Bruckner, S., Hüffner, F., Karp, R.M., Shamir, R. & Sharan, R. 2009. Topology-free querying of protein interaction networks. *In Proceedings of the 13th Annual International Conference on Research in Computational Molecular Biology (RECOMB '09), Tucson, Arizona, USA. May 2009. Vol. 5541 in Lecture Notes in Bioinformatics, Springer.*

Bunke, H. & Riesen, K. 2008. Graph Classification Based on Dissimilarity Space Embedding. *Lecture Notes in Computer Science 5342: 996–1007.*

Caetano, T.S., McAuley, J.J., Cheng, L., Le, Q.V., Smola, A.J. 2009. Learning graph matching. *IEEE Trans Pattern Anal Mach Intell. 31(6): 1048–58.*

Cao, Y., Jiang, T., Girke, T. 2008. A maximum common substructure-based algorithm for searching and predicting drug-like compounds. *Bioinformatics 24(13): i366–374.*

Capriotti, E. & Marti-Renom, M.A. 2008. Computational RNA structure prediction. *Current Bioinformatics 3: 32–45.*

Carvalho, L.E., Lawrence, C.E. 2008. Centroid estimation in discrete high-dimensional spaces with applications in biology. *Proc Natl Acad Sci U S A. 105(9): 3209–3214.*

Catana, C. 2009. Simple Idea to Generate Fragment and Pharmacophore Descriptors and Their Implications in Chemical Informatics. *Journal of Chemical Information and Modeling 49(3): 543–548.*

Chao, S-Y. 2008. Maximum Common Substructure Extraction in RNA Secondary Structures Using Clique Detection Approach. *Proceedings Of World Academy Of Science, Engineering And Technology 35: 219–228.*

Cho, J., Choi, J., Qiao, M., Ji, C.W., Kim, H.-Y., Uhm, K.-B. & Chon, T.-S. 2007. Automatic identification of whiteflies, aphids and thrips in greenhouse based on image analysis. *International Journal Of Mathematics And Computers In Simulation 1(1): 46–53.*

Chu, C-H., Tang, C.Y., Tang, C.-Y. & Pai, T-W. 2008. Angle-distance image matching techniques for protein structure comparison. *Journal of Molecular Recognition 21(6): 442–452.*

Cline, M.S., Smoot, M., Cerami, E., Kuchinsky, A., Landys, N., Workman, C., Christmas, R., Avila-Campilo, I., Creech, M., Gross, B., Hanspers, K., Isserlin, R., Kelley, R., Killcoyne, S., Lotia, S., Maere, S., Morris, J., Ono, K., Pavlovic, V., Pico, A.R., Vailaya, A., Wang, P.L., Adler, A., Conklin, B.R., Hood ,L., Kuiper, M., Sander, C., Schmulevich, I., Schwikowski, B., Warner, G.J., Ideker, T., & Bader, G.D. 2007. Integration of biological networks and gene expression data using Cytoscape. *Nature Protocols 2(10): 2366–2382.*

Cozzuto, L., Petrillo, M., Silvestro, G., Di Nocera, P.P. & Paolella, G. 2008. Systematic identification of stem-loop containing sequence families in bacterial genomes. *BMC Genomics 9: 20.*

Crabtree, J.D. & Mehta, D.P. 2009. Automated reaction mapping. *Journal of Experimental Algorithmics 13: 1.15–1.29.*

De Gregorio, E., Bertocco, T., Silvestro, G., Carlomagno, M.S., Zarrilli, R. & Di Nocera, P.P. 2009. Structural organization of a complex family of palindromic repeats in Enterococci. *FEMS Microbiol Lett. 292(1): 7–12.*

Faulon, J.L. *et al.* 2004. The signature molecular descriptor. 4. Canonizing molecules using extended valence sequences. *J. Chem. Inf. Comput. Sci. 44: 427–436.*

Faulon, J.L., Misra, M., Martin, S., Sale, K., Sapra, R. 2008. Genome scale enzyme-metabolite and drug-target interaction predictions using the signature molecular descriptor. *Bioinformatics 24(2): 225–233.*

Fernandes, M.B., Scotti, M.T., Ferreira, M.J., Emerenciano, V.P. 2008. Use of self-organizing maps and molecular descriptors to predict the cytotoxic activity of sesquiterpene lactones. *European Journal of Medicinal Chemistry 43(10): 2197–2205.*

Ferro, A., Giugno, R., Mongiovì, M., Pulvirenti, A., Skripin, D. & Shasha, D. 2008. GraphFind: enhancing graph searching by low support data mining techniques. *BMC Bioinformatics. 9(Suppl 4): S10.*

Gan, H.H., Fera, D., Zorn, J., Shiffeldrim, N., Tang, M., Laserson, U., Kim, N. & Schlick, T. 2004. RAG: RNA-As-Graphs database--concepts, analysis, and features. *Bioinformatics 20: 1285–1291.*

Giugno R & Shasha D 2002. GraphGrep: A Fast and Universal Method for Querying Graphs. *Proceedings of the 16th International Conference on Pattern Recognition: 11–15 August 2002; Quebec: 112–115.*

Hamada, M., Kiryu, H., Sato, K., Mituyama, T., Asai, K. 2009 Prediction of RNA secondary structure using generalized centroid estimators. *Bioinformatics 25(4): 465–473.*

Han, J., Cheng, H., Xin, D. & Yan, X. 2007. Frequent pattern mining: current status and future directions. *Data Mining and Knowledge Discovery 15(1): 55–86.*

Han, K., Ju, B.H. 2003. A fast layout algorithm for protein interaction networks. *Bioinformatics 19(15): 1882–1888.*

Han, K., Ju, B.H., Jung, H. 2004. WebInterViewer: visualizing and analyzing molecular interaction networks. *Nucleic Acids Res. 32(Web Server issue): W89–95.*

Havukkala, I., Benuskova, L., Pang, P., Jain, V., Kroon, R. & Kasabov, N. 2005. Image and fractal information processing for large-scale chemoinformatics, genomics analyses and pattern discovery. *In: Lecture Notes in Bioinformatics Eds. J.C. Rajapakse, L. Wong & R. Acharya, Spinger Verlag, LNBI 4146: 163–173.*

Havukkala, I., Pang, S. N., Jain, V. & Kasabov, N. 2005. Classifying MicroRNAs by Gabor Filter Features from 2D Structure Bitmap Images on a Case Study of Human microRNAs. *Journal of Computational and Theoretical Nanotechnology 2(4): 506–513.*

Höchsmann, M., Voss, B., Giegerich, R. 2004. Pure Multiple RNA Secondary Structure Alignments: A Progressive Profile Approach. *IEEE/ACM Transactions on Computational Biology and Bioinformatics 1: 53–62.*

Hsu, P.W., Huang, H.D., Hsu, S.D., Lin, L.Z., Tsou, A.P., Tseng, C.P., Stadler, P.F., Washietl, S. & Hofacker, I.L. 2006. miRNAMap: genomic maps of microRNA genes and their target genes in mammalian genomes. *Nucleic Acids Res. 34(Database issue): D135–139.*

Hu, Z., Snitkin, E.S. & DeLisi, C. 2008. VisANT: an integrative framework for networks in systems biology. *Brief Bioinform. 9(4): 317–325.*

Hur, J. & Wild, D.J. 2008. PubChemSR: A search and retrieval tool for PubChem. *Chem Cent J. 2: 11.*

Ilonen, J.; Kamarainen, J.-K. & Kalviainen, H. 2007. Fast extraction of multi-resolution Gabor features. *14th International Conference on Image Analysis and Processing ICIAP 2007, 10–14 Sept. pp. 481–486.*

Jacob, L., Hoffmann, B., Stoven, V., Vert, J.P. 2008. Virtual screening of GPCRs: an in silico chemogenomics approach. *BMC Bioinformatics 9: 363.*

Jacob, L., Vert, J.P. 2008. Protein-ligand interaction prediction: an improved chemogenomics approach. *Bioinformatics 24(19): 2149–2156.*

Jakuschev, S. & Hoffmann, D. 2009. A Novel Algorithm for Macromolecular Epitope Matching. *Algorithms 2: 498–517.*

Jalba, A.C., Wilkinson, M.H.F., Roerdink, J.B.T.M., Bayer, M.M. & Juggins, S. 2008. Automatic diatom identification using contour analysis by morphological curvature scale spaces *Machine Vision and Applications 16(4): 217–228.*

Jónsdóttir, S.O., Jørgensen, F.S. & Brunak, S. 2005. Prediction methods and databases within chemoinformatics: emphasis on drugs and drug candidates. *Bioinformatics. 21(10): 2145–2160.*

Kamarainen, J.K., Kyrki, V. & Kälviäinen, H. 2006. Invariance properties of gabor filter-based features--overview and applications. *IEEE Trans Image Process. 15(5): 1088–1099.*

Kim, S.-W. & Gao, J. 2008. On Using Dimensionality Reduction Schemes to Optimize Dissimilarity-Based Classifiers. *Lecture Notes in Computer Science 5197: 309–316.*

Knisley, D. & Knisley, J. 2008. Graph theoretic models in chemistry and molecular biology. *Chapter 3 In: Nayak, A. & Stojmenovic, I. (eds) Handbook of applied algorithms. Wiley-IEEE, 544 pp.*

Kuijper, A., Olsen, O. F:, Giblin, P. J. & Nielsen, M. 2006. Alternative 2D Shape Representations using the Symmetry Set. *Journal of Mathematical Imaging and Vision 26(1/2): 127–147.*

Kuijper, A. & Havukkala, I. 2007. Computationally Efficient Matching of MicroRNA Shapes using Mutual Symmetry. *9th IASTED International Conference on Signal and Image Processing (SIP 2007, Honolulu, Hawaii, USA, August 20–22, 2007), pages 477–482.*

Le, Q., Pollastri, G. & Koehl, P. 2009. Structural alphabets for protein structure classification: a comparison study. *J Mol Biol. 387(2): 431–50.*

Li, K., Rahman, R., Gupta, A., Siddavatam, P. & Gribskov, M. 2008. Pattern Matching in RNA Structures. *Lecture Notes in Computer Science 4983: 317–330.*

Li, W. 2006. A fast clustering algorithm for analyzing highly similar compounds of very large libraries. *J Chem Inf Model. 46(5): 1919–1923.*

Liu, J., Wang, J.T., Hu, J., Tian, B. 2005. A method for aligning RNA secondary structures and its application to RNA motif detection. *BMC Bioinformatics. 6: 89.*

Lo, W.C. & Lyu, P.C. 2008. CPSARST: an efficient circular permutation search tool applied to the detection of novel protein structural relationships. *Genome Biol. 9(1): R11.*

Lo, W.C., Huang, P.J., Chang, C.H., Lyu, P.C. 2007. Protein structural similarity search by Ramachandran codes. *BMC Bioinformatics 8: 307.*

Malod-Dognin, N., Andonov, R. & Yanev, N. 2009. Solving Maximum Clique Problem for Protein Structure Similarity. *ArXiv e-prints, 2009arXiv 0901.4833M*

Margraf, T. & Torda, A. 2008. HANSWURST: Fast Efficient Multiple Protein Structure Alignments. *In: Proceedings of the NIC Workshop 2008 Computational Biophysics to Systems Biology (CBSB08), Ulrich H. E. Hansmann, Jan H. Meinke, Sandipan Mohanty, Walter Nadler, Olav Zimmermann (Editors), NIC Series, 40: 313–316.*

Marsden, R.L., Lee, D., Maibaum, M., Yeats, C. & Orengo, C.A. 2006. Comprehensive genome analysis of 203 genomes provides structural genomics with new insights into protein family space. *Nucleic Acids Res. 34(3): 1066–1080.*

Mayo, M. & Watson, A.T. 2006. Automatic Species Identification of Live Moths. *Pp. 46–59 in: Applications and Innovations in Intelligent Systems XIV: Proceedings of AI-2006, the Twenty-Sixth SGAI International Conference on Innovative Techniques and Applications of Artificial Intelligence, Eds. Richard Ellis, Tony Allen, Andrew L. Tuson, British Computer Society. Specialist Group on Artificial Intelligence. Springer.*

Mingqiang, Y., Kidiyo, K. & Joseph, R. 2008. A Survey of Shape Feature Extraction Techniques. *Pp. 44–90. In: Peng-Yeng Yin (ed), Pattern Recognition Techniques, Technology and Applications, I-Tech, Vienna, Austria.*

Mokhtarian, F. & Bober, M. Z. 2003. Curvature Scale Space Representation: Theory, Applications, & Mpeg-7 Standardization. *Kluwer Academic Publishers, Dordrecht, The Netherlands.*

O'Neill, M.A. 2007. DAISY: A Practical Computer Based Tool for Semi Automated Species Identification. *In: Automated Taxon Identification in Systematics, Theory Approaches and Applications, N. MacLeod (Ed.) CRC Press, Ch. 7, pp101–114.*

Otto, W., Will, S. & Backofen, R. 2008. Structure Local Multiple Alignment of RNA. *Lecture Notes in Informatics P-136, 178–188.*

Pavlopoulos, G. Gap, A., Wegener, A., Aw, L. & Schneider, R. 2008. A survey of visualization tools for biological network analysis. *BioData Min. 1(1):12.*

Peng, H. 2008. Bioimage informatics: a new area of engineering biology. *Bioinformatics. 24(17): 1827–1836.*

Pieper, U., Eswar, N., Webb, B.M., Eramian, D., Kelly, L., Barkan, D.T., Carter, H., Mankoo, P., Karchin, R., Marti-Renom, M.A., Davis, F.P. & Sali, A. 2009. MODBASE, a database of annotated comparative protein structure models and associated resources. *Nucleic Acids Res. 37(Database issue): D347–354.*

Raymond, J.W. & Willett, P. 2002. Maximum common subgraph /isomorphism algorithms for the matching of chemical structures. *Journal of Computer-aided Molecular Design, 16: 521–533.*

Riesen, K. & Bunke, H. 2009. Reducing the dimensionality of dissimilarity space embedding graph kernels. *Engineering Applications of Artificial Intelligence 22(1): 48–56.*

Rosselló, F. & Valiente, G. 2005. Graph Transformation in Molecular Biology. *Lecture Notes in Computer Science 3393: 116–133.*

Sam, V., Tai, C.H., Garnier, J., Gibrat, J.F., Lee, B. & Munson, P.J. 2008. Towards an automatic classification of protein structural domains based on structural similarity. *BMC Bioinformatics 9: 74.*

Samson, A.O., Levitt, M. 2009. Protein segment finder: an online search engine for segment motifs in the PDB. *Nucleic Acids Res. 37(Database issue): D224–228.*

Sankoff, D. 1985. Simultaneous solution of the RNA folding, alignment and protosequence problems. *SIAM Journal on Applied Mathematics 1985; 45: 810–825.*

Sebastian, T.B., Klein, P.N. & Kimia, B.B. 2004. Recognition of shapes by editing their shock graphs. *IEEE Trans Pattern Anal Mach Intell. 26(5): 550–571.*

Shu, W., Bo, X., Zheng, Z. & Wang, S. 2008, A novel representation of RNA secondary structure based on element-contact graphs. *BMC Bioinformatics 9: 188.*

Siebert, S. & Backofen, R. MARNA: multiple alignment and consensus structure prediction of RNAs based on sequence structure comparisons. *Bioinformatics 21(16): 3352–3359.*

Suderman, M. & Hallett, M. 2007. Tools for visually exploring biological networks. *Bioinformatics 23(20): 2651–2659.*

Sun, B., Mitra, P. & Giles, C.L. 2008. Mining, Indexing, and Searching for Textual Chemical Molecule Information on the Web. *The 17th International World Wide Web Conference (WWW'08), Beijing, China.*

Swamidass, S.J. *et al.* 2005. Kernels for small molecules and the prediction of mutagenicity, toxicity and anti-cancer activity. *Bioinformatics 21(Suppl. 1): i359–i368.*

Tangelder, J.W.H. & Veltkamp, R.C. 2008. A survey of content based 3{D} shape retrieval methods. *Multimedia Tools and Applications, 39, 441–471.*

Tian, Y., McEachin, R.C., Santos, C., States, D.J. & Patel, J.M. 2007. SAGA: a subgraph matching tool for biological graphs. *Bioinformatics 23(2): 232–239.*

Trepalin, S.V. & Yarkov, A.V. 2008. Hierarchical Clustering of Large Databases and Classification of Antibiotics at High Noise Levels. *Algorithms 1: 183–200.*

Vert, J.P. & Jacob, L. 2008. Machine learning for in silico virtual screening and chemical genomics: new strategies. *Comb Chem High Throughput Screen. 11(8): 677–85.*

Vitkup, D., Melamud, E., Moult, J. & Sander, C. 2001. Completeness in structural genomics. *Nature Struct. Biol. 8: 559–566.*

Wang, X., Smalter, A., Huan, J. & Lushington, G. H. 2009. G-hash: towards fast kernel-based similarity search in large graph databases. *In: Proceedings of the 12th international Conference on Extending Database Technology: Advances in Database Technology (Saint Petersburg, Russia, March 24–26, 2009). M. Kersten, B. Novikov, J. Teubner, V. Polutin, and S. Manegold, Eds. EDBT '09, vol. 360. ACM, New York, NY, 472–480.*

Washio , T. & Motoda, H. 2003. State of the art of graph-based data mining. *ACM SICKDD Explorations Newsletter, 5(1): 59–68.*

Wei, L, Keogh, E., Xi, X & Yoder, M. 2008.Efficiently finding unusual shapes in large image databases. *Data Mining and Knowledge Discovery 17(3): 343–376.*

Weis, D.C., Visco, D.P.Jr. & Faulon, J.L. 2008. Data mining PubChem using a support vector machine with the Signature molecular descriptor: classification of factor XIa inhibitors. *J Mol Graph Model. 27(4): 466–475.*

Will, S., Reiche, K., Hofacker, I.L., Stadler, P.F. & Backofen, R. 2007. Inferring noncoding RNA families and classes by means of genome-scale structure-based clustering. *PLoS Comput Biol 3(4): e65.*

Yadav, M.K., Kelley, B.P. & Silverman, S.M. 2004. The Potential of a Chemical Graph Transformation System. *Lecture Notes in Computer Science 3256: 83–95.*

Yankov, D., Keogh, E., Wei, L., Xi, X. & Hodges, W. 2008. Fast Best-Match Shape Searching in Rotation-Invariant Metric Spaces. *IEEE Transactions on Multimedia 10(2): 230–239.*

Zaki, M. J. 2005. Efficiently Mining Frequent Trees in a Forest: Algorithms and Applications. *IEEE Trans. Knowl. Data Eng. 17(8): 1021–1035.*

Zhang, D. & Lu, G. 2004. Review of shape representation and description techniques. *Pattern Recognition 37(1): 1–19.*

Chapter 6

Function Annotation and Ontology Based Searching and Classification

One of the most important tasks in making genomic data more useful is to characterize the new gene and protein sequences based on known functions of similar genes and proteins, *e.g.* annotation using "guilt by association". Similarity may be based on overall sequence similarity, conserved DNA or protein motifs, location to nearby genes of known function, correlation of expression profile in time or place to other known genes and so on. *In silico* annotation adds enormously to the available information, because biological experiments for gene function verification are time-consuming and expensive. Thus already a majority of the gene function labels in current databases are based on *in silico* predictions from a much smaller set of annotations from well-studied model organisms like mouse, zebrafish, banana fly, yeast *Saccharomyces cerevisiae*, bacterium *Escherichia coli etc.*

6.1 Annotation ontologies

For function annotation, a standardized nomenclature with semantic relations between terms, i.e. an ontology, for various biological activities of a gene or protein is needed, and such normalized semantic ontologies are already in widespread use. Foremost among them is the well-known Gene Ontology (www.geneontology.org), with a specially developed databrowser called AmiGO (Carbon *et al.* 2009). This big ontology is linked to many other ontologies that are being built as part of the standardization and generation of metadata in biological sciences. The building of such Semantic Web for biology includes creation and linking of metadata using standard controlled vocabularies and ontologies that can be queried and analyzed by humans and computers alike.

An example of a hierarchy of terms for one gene product in Gene Ontology is shown in Figure 6-1. Many other similar bio-ontologies are being built and implemented, for example protein motif and related annotation features have recently been standardized into the Protein Feature Ontology (Reeves *et al.* 2008) providing structured controlled vocabulary of 140 concepts for protein sequence and structure features. It has been integrated into the Sequence Ontology (SO) in the clearinghouse for biological ontologies, called the Open Biomedical Ontologies consortium (obofoundry.org). The Open Biomedical Ontologies website already contains over 50 ontologies ranging from amphibian gross anatomy to mouse pathology to pathogen transmission *etc*.

Gene Ontology is the widely adopted function nomenclature standard, but many others are used as well. The established E.C.

⊟ ▯ GO:0008150 : biological_process [168302 gene products]
 ⊟ ▯ GO:0065007 : biological regulation [32662 gene products]
  ⊟ ▯ GO:0050789 : regulation of biological process [29954 gene products]
   ⊟ ▯ GO:0048518 : positive regulation of biological process [7802 gene products]
    ⊟ ▯ GO:0048522 : positive regulation of cellular process [5069 gene products]
     ⊟ ▯ GO:0031325 : positive regulation of cellular metabolic process [2556 gene products]
      ⊟ ▯ GO:0031328 : positive regulation of cellular biosynthetic process [2158 gene products]
       ⊟ ▯ **GO:0045727 : positive regulation of translation [71 gene products]**
        ⊟ ▯ GO:0070131 : positive regulation of mitochondrial translation [1 gene product]
        ⊟ ▯ GO:0046012 : positive regulation of oskar mRNA translation [4 gene products]
        ⊟ ▯ GO:0032056 : positive regulation of translation in response to stress [1 gene product]
        ⊟ ▯ GO:0045975 : positive regulation of translation, ncRNA-mediated [2 gene products]
        ⊟ ▯ GO:0045901 : positive regulation of translational elongation [0 gene products]
        ⊟ ▯ GO:0045903 : positive regulation of translational fidelity [5 gene products]
        ⊟ ▯ GO:0045948 : positive regulation of translational initiation [18 gene products]
         ⊟ ▯ **GO:0070134 : positive regulation of mitochondrial translational initiation** [1 gene product]

Figure 6-1. An example of Gene Ontology biological process concept hierarchy, with a final link to one gene product on last line. The AmiGO Gene Ontology browser view of the same gene product is shown in Figure 6-2.

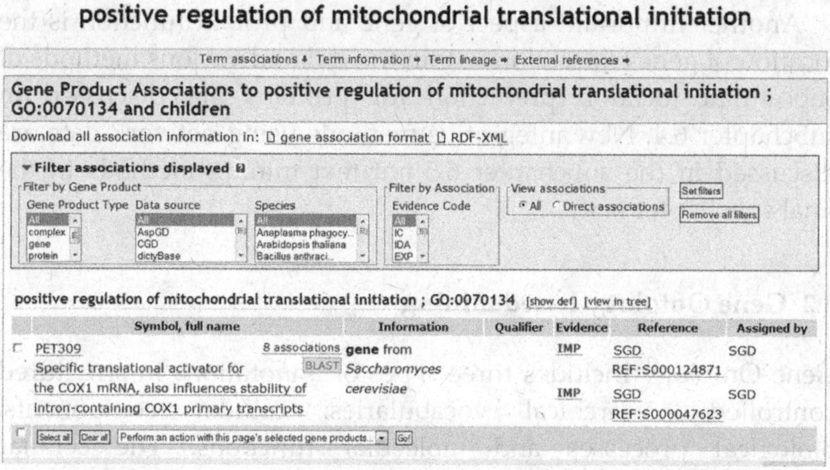

positive regulation of mitochondrial translational initiation

Term associations ▲   Term information ▼   Term lineage ▼   External references ▼

Gene Product Associations to positive regulation of mitochondrial translational initiation ; GO:0070134 and children

Download all association information in: 🗋 gene association format 🗋 RDF-XML

▼ **Filter associations displayed** ⚑

Filter by Gene Product			Filter by Association	View associations	
Gene Product Type	Data source	Species	Evidence Code	⦿ All ⦾ Direct associations	Set filters
All	All	All	All		Remove all filters
complex	AspGD	Anaplasma phagocy...	IC		
gene	CGD	Arabidopsis thaliana	IDA		
protein	dictyBase	Bacillus anthraci...	EXP		

positive regulation of mitochondrial translational initiation ; GO:0070134 [show def] [view in tree]

	Symbol, full name		Information	Qualifier	Evidence	Reference	Assigned by
☐	PET309	8 associations	**gene** from		IMP	SGD	SGD
	Specific translational activator for	BLAST	Saccharomyces			REF:S000124871	
	the COX1 mRNA, also influences stability of		cerevisiae		IMP	SGD	SGD
	intron-containing COX1 primary transcripts					REF:S000047623	

☐ [Select all] [Clear all] [Perform an action with this page's selected gene products...▾] [Go!]

Figure 6-2. AmiGO browser view of the gene product for the biological process of positive regulation of mitochondrial translational initiation in the Gene Ontology. The hyperlink to PET309 gene leads to SwissProt and further databases.

(Enzyme Commission) numbering system for enzyme-catalyzed reactions is a hierarchical number code. For example, EC 2.4 is for glycosyl transferases, EC 2.4.2 for pentosyl transferases, EC 2.4.2.15 guanosine phosphorylase, also known with its systematic name guanosine:phosphate α-D-ribosyltransferase, and so on. The Gene Ontology links to EC numbers, as well as to many other databases, including EMBL, SwissProt and InterPro *etc.* InterPro protein motif database also has functional labels, as well as links to many other motif databases.

Function annotation is thus a complex affair, depending on what kind of function labels are to be assigned to the gene or protein to be annotated. Currently the main function annotation systems try to use the Gene Ontology functions for already annotated sequences to label new sequences, as discussed in the next subchapter. The other main approach is based on simple sequence similarity or other existing genome annotations for model organisms, discussed in subchapter 6.3.

Another important aspect of gene and protein function is the location of gene expression within the cell and various methods of subcellular location prediction for proteins are discussed in subchapter 6.4. New integrative methods using network data are discussed in the subchapter 6.5 ant text mining methods in the final subchapter 6.6.

6.2 Gene Ontology based mining

Gene Ontology includes three types of annotations in structured controlled hierarchical vocabularies: cellular components, biological processes and molecular functions. The cellular components vocabulary identifies the location in which the gene or gene product (RNA or protein) is active, for example nucleus, mitochondrion and cell membrane. The biological process indicates in which multistep process the gene or gene product takes part in, for example, DNA synthesis, sugar transport, RNA translation to protein and so on. The molecular function tells what the molecular level activity is, for example, protein phosphorylation, a specific enzyme activity, receptor binding.

Many tools exist for GO data processing. A total of 68 Gene Ontology based gene list analysis tools were reviewed by Huang *et al.* (2008a). Such enrichment analysis based tools can be divided into three main groups. The first group is the traditional approach, in which the P-value for enrichment is calculated for each Gene Ontology term allocated to the inputted pre-selected gene list, as compared to a random selection from the standard gene set. These algorithms include the popular FatiGO and GoMiner.

The second group is based on analysing the whole dataset of genes which are ranked in order, without selecting *e.g.* a subset of most significantly changed (over- or under-expressed) genes from the microarray. This uses more information and should lead to more sensitive detection of biological functions. These algorithms include GSEA and FatiScan. Both the first and second group methods are more sensitive to the larger changes in gene expression values, which lead to the top-ranking genes, but the smaller, but biologically significant changes in crucial regulatory genes may be equally important, but are not as easily detected by these methods. Another disadvantage is that often more complex multivariate data sets need to be analyzed, not just simple comparison of two datasets.

The third group method is based on the analysis of the network connections between the GO terms, adding more information derived from the GO network. TopGO (Alexa *et al.* 2006) and ProfCom (Antonov *et al.* 2008) are examples of this group. The multitude of software and approaches taken to calculate enrichment probabilities is an indication of the multifaceted interests of bioinformaticians and the relative novelty of the biological ontologies.

There are also many other ways to calculate the similarities of proteins or protein sets using different semantic distance measures. For example, Chagoyen *et al.* (2008) used a measure of coherence, which is defined as the sum of averaged pairwise cosine similarities between all pairs of proteins in the target set of proteins, normalized to between 0 and 1.

Many other similarity scoring systems are also possible (see *e.g.* Wang *et al.* 2007, Lerman & Shakhnovich 2007), but have not been compared side by side yet for their performance. Mistry (2008) proposed a simple and fast term overlap measure and Del Pozo *et al.* (2008) proposed a novel method based on calculating the co-occurrence matrix of genes and GO labels, then calculating similarity matrix based on cosine distance, reducing the dimensionality of the matrix by spectral clustering and finally clustering the data using hierarchical clustering. This is reminiscent of the Latent Semantic Indexing discussed in a later subchapter 9.8. A solid comparison for the many methods for the statistical arguments and the analysis workflows by a good benchmark is needed.

A few other recent Gene Ontology tools are worth a special mention in this active research area. An algorithm named

COFECO integrates KEGG data and GO data to display metabolic and biological process network diagrams for specific gene lists in a convenient web interface, and is a good example of modern integrative biodata analysis (Sun *et al.* 2009). CORNA (Wu & Watson 2009) is R-package based software which checks a given gene list for enrichment for microRNA target associations from mirBase data. Its accuracy is based on the mirBase database contents, so that it cannot be said to be a truly *ab initio* analysis, but it still usefully leverages existing data on known gene sets regulated by microRNAs.

Another very recent tool of this kind is GeneSet2miRNA (Antonov *et al.* 2009), which uses the predicted microRNA targets from another new microRNA database, miRecords (Xiao *et al.* 2008) as a basis for analysing an inputted gene list for enrichment. The predicted targets from miRecords database were obtained using 11 different target prediction algorithms, and each target was considered the more reliable, the more of the 11 algorithms agreed with the target designation. Because searching all possible combinations of targets to find the best one that matches the input list is not feasible, a greedy heuristics is used in Geneset2miRNA, and p-values for multiple hypothesis testing are adjusted by a Monte-Carlo simulation, assuming that the genes in the input list were sampled from the whole set of genes in the genome. This assumption is not always correct, because often a microarray used to obtain data does not contain all the expressed genes in the genome that are available in the miRecords database. Therefore caution is needed to interpret the significance values, which may be artificially high due to this bias. The adjustment of probability values for multiple testing in this kind of algorithms is discussed in detail by Wu & Watson (2009).

One of the best current tools for analysis a gene list of interest is GOrilla (Eden *et al.* 2009). Its advantages include those of fast speed, settable probability threshold and graphical network representation of results. Other similar well performing flexible tools are Fatiscan and GO-stat. As often is the case, it is advisable to use two or three different tools to compare results for detecting outliers or items not found by one specific approach.

In view of the multitude of available tools, a consolidation to a few commonly adopted algorithms would be beneficial, so that improvement efforts could be more efficiently focused on the best tools. Perhaps better communication and standardized comparisons of the various software using common benchmarks would help in this, because standardized benchmarks have been very useful in the data mining and artificial intelligence research within the engineering and knowledge engineering sciences.

6.3 Sequence similarity based function prediction

Sequence similarity based data mining is the main method of function prediction which is based on prior annotation of similar genes and proteins. This is a very general approach, and guilt by association is indeed a powerful method to add *in silico* derived information to large numbers of genes and proteins, all of which cannot be studied experimentally. Thus the first step in standard new genome analysis is to compare the new data to known genomes and their functional annotations to get a first pass idea of what the new genome contains. Most major genome repositories run routinely their continuously improved annotation engines against the new genomes being added at increasing pace.

Such methods have become so powerful, that whole genome annotations are already done almost automatically, especially in bacterial genomics, using sophisticated software packages which do automatic blasting of predicted gene and protein sequences to known genomes and automatic linking to the annotations of the matching genes to the new genome gene candidates. The multitude of software already available is reviewed by Médigue & Moszer (2007).

Notable recent systems include PUMA2 (Maltsev *et al.* 2006) and BASys (Van Domselaar *et al.* 2005, Stothard & Wishart 2006). PIPA (Yu *et al.* 2008) is an automated proteome annotation pipeline and MED (Zhu *et al.* 2007) predicts reliably in an unsupervised manner bacterial and archaeal genes to form a solid basis for protein translation and subsequent proteome annotation. Another bacteria-focused visualization method does automated blasting for orthologous genes of bacterial genomes and shows the genomes with colour-coded matching orthologs on aligned circles in a zoomable image (Grant & Stothard 2008). For metagenomes, a good automated web-accessible annotation pipeline called RAST Server is already in ample use, speeded up by grid-based distributed computing resources (Meyer *et al.* 2008).

A notable new approach is to utilize also phylogenetic profiles in addition to consensus gene functions from known genomes, in order to enhance gene function prediction quality. In bacteria such a system is working well (Lin *et al.* 2009), based on simple blasting to known genomes for gene orthologs as a starting point. Both statistical enrichment analysis of related genes in the pathway of the analysed gene as well as an inductive method of using training sets for various supervised learning algorithms (like SVM, Naïve Bayes, logistic regression *etc.*, with 10-fold crossvalidation using the WEKA package) were used to determine gene function with good accuracy.

The best new systems perform extensive BLAST similarity searches combined with clustering and chaining of relationships in Gene Ontology. The PFP system (Hawkins *et al.* 2009) assigns Gene Ontology functions to 60–90% of new sequences with about 80% accuracy, which is significantly better than plain PSI-BLAST on individual sequences or InterPro motif search by InterProScan. Recently this system was improved to an ESG (Extended Similarity Grouping) system, in which each Gene Ontology function is given a probability value which is refined by iterative PSI-BLAST searches (Chitale *et al.* 2009). The method appears superior to using top BLAST hits or the previous state of art PFP method. The ESG method could be further improved by including information about the correlations of terms within the Gene Ontology network of semantic relationships.

6.4 Cellular location prediction

The localization of the protein within a cell is an important indicator of its function or involvement in specific metabolic pathways, *e.g.* energy metabolism in the mitochondrion or photosynthesis in chloroplasts, and therefore algorithms predicting the subcellular localization of proteins are an important value-adding annotation to new unannotated proteins. There has been already a long history of developing such methods, which can be divided into three main groups, as reviewed by Casadio *et al.* (2008).

Firstly, sequence similarity based methods use proteins with known cellular localization to infer the putative location of the new unknown proteins in the cell. This is a powerful method, but can give erroneous results in cases where the protein has changed its targeting during evolution or where the localization signal is weak or where there are few known localizations for the related sequences in the database. After simple blasting for the orthologous genes, one can then use other information for all the sequences for SVM or Bayesian inference to classify the new sequence for its localization. Whole genomes have been annotated by this approach in the database eSLDB (Pierleoni *et al.* 2007), which contains in addition to the homology-based classification also experimental data and predicted localization using three different methods.

Secondly, specific sequence motifs or regional protein domain characteristics can be searched and compared to known proteins using SVM, Bayesian networks and so on. The motif databases PFAM, SMART and InterPro contain several motifs related to cellular localization, which can be used together with other computed features from the sequences for machine learning methods of classification. These methods include pTARGET, Domain Projection and PSLT2.

Thirdly, *ab initio* analysis for protein localization based on amino acid sequence only has been a dominant strategy, and a large number of algorithms have been constructed during the past

10 years, utilizing various computed features and SVM, HMM, neural networks and so on (for a long listing, see www.psort.org).

Many of the earlier methods focused on the so-called signal peptide characteristics, because the leading segment in the protein often encodes special instructions for the cell machinery for transporting the protein to a desired location, especially to excretion from cell. This signal peptide is normally a dozen or so amino acids long specific sequence, often conserved in closely related organisms. For targeting to cell membranes, the amino acids have a specific hydrophobicity profile that can be used to predict proteins of this class. However, these methods have not been shown to outperform simple sequence similarity based assignments, and may in some cases perform much worse (Sprenger *et al.* 2006).

Some methods specifically look for signal peptides, nucleus localization signals or transit peptides, but are not considered to outperform more general methods for multiclass prediction. However, the use of SVM using amino acid composition for cellular location prediction was pioneered successfully by Tamura & Akutsu (2007); their method is called SLPFA (for Subcellular Location Prediction with Frequency and Alignment).

The top three current algorithms, which all belong to the third group of methods, can be said to be Wolf PSORT (Horton *et al.* 2007), BaCelLo (Pierleoni *et al.* 2006) and LocTree (Nair & Rost 2005). The first two methods perform best for animal and fungal proteins, and the last two are best for plant proteins. WolfPSORT is a rule-based decision tree system using N-terminal sequence, amino acid composition and empirical rules. BaCelLo uses SVM decision trees, amino acid composition and profiles and N- and C-terminal sequences. LocTree uses SVM decision trees with amino acid profiles, secondary structure and N-terminal sequence.

Notably, Wolf PSORT includes 12 cellular locations to be predicted, while BaCelLo and LocTree only predict 5 and 6 classes, respectively. Only Wolf PSORT can predict a protein to be simultaneously targeted to two different locations, which certainly is the case for some bifunctional proteins. One remaining problem for these algorithms is to predict which kind of membrane are the TM proteins targeted to (endoplasmic reticulum, mitochondrion, nuclear membrane *etc.*). It is unknown whether the amino acid sequence alone includes sufficient information for this kind of classification; in addition, paucity of experimental data and lack of knowledge of the transporting systems is also a problem for building good learning and testing datasets for specific membrane targeting.

For special groups of organisms, generic methods will not always work well, so that specific taxon-tuned algorithms need to be developed. An interesting example is that of signal peptides in heterokont organisms (Figure 6-3) which include diatoms, brown algae and plant pathogenic *Oomycetes* fungi (*e.g.* potato blight) and which evolved as a fusion of an alga within a single-celled eukaryote. In these complex eukaryotic single-celled microbes the targeting of proteins to innermost chloroplast needs passing through two sets of membranes (Gschloessl *et al.* 2008). Their heterokont-specific algorithm includes hierarchical SVMs and reached an overall accuracy of 96.3%. It also performs well for two types of signal peptides from other eukaryotes, and is thus a recent competitor to the above-mentioned Wolf PSORT. A comparative evaluation is needed, and perhaps the two methods could be merged for even better performance.

Another recent impressive cellular location prediction system is CellPloc (Chou & Shen 2008), which uses a comprehensive dataset compiled from SwissProt and utilizes also Gene Ontology annotation as well specific *ab initio* algorithms tuned to different groups of organisms (plants, viruses, human *etc.*).

6.5 New integrative methods: Utilizing networks

The trend to integrative biology in general is also noticeable in gene function prediction technologies in bioinformatics. One forerunner of the modern integrative methods was Global Mapping of Unknown Proteins (GMUP) system (Xiong *et al.* 2006),

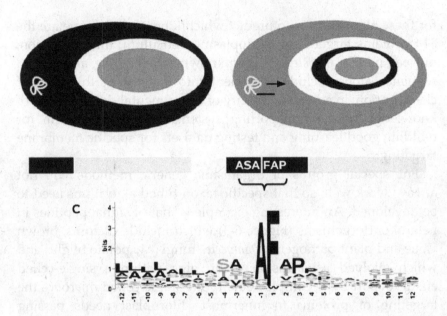

Figure 6-3. Left: Schematic of an algal cell with a chloroplast within, totaling 2 internal membranes. Right: A heterokont cell with an algal "cell" within, totaling 4 internal membranes. The heterokont cell needs two targeting peptide motifs (ASA signal peptide and FAP chloroplast targeting peptide) for the protein to reach the innermost chloroplast. The sequence logo of the heterokont targeting domain is shown at the bottom. From Gschloessl (2008). Reproduced by Creative Commons Attribution License from an Open Access article.

which first extracted Gene Ontology annotations based on BLAST similarity derived protein pairs and then used these annotations as an input to a neural network. A total of 16 different data sources (*e.g.* Rosetta compendium expression data, MIPS gene interaction data, BIND protein complex data and so on) were used to derive the annotations, and the number of evidence from each data source used as a feature in the feature vector of 16 numbers. The construction of the feature vector is shown in Figure 6-4.

The feature vectors were processed by a three layer back propagation neural network with an input layer of 16 nodes and a hidden layer of 8 nodes and 10% of the data was used for crossvalidation training sets. The system was shown to be very

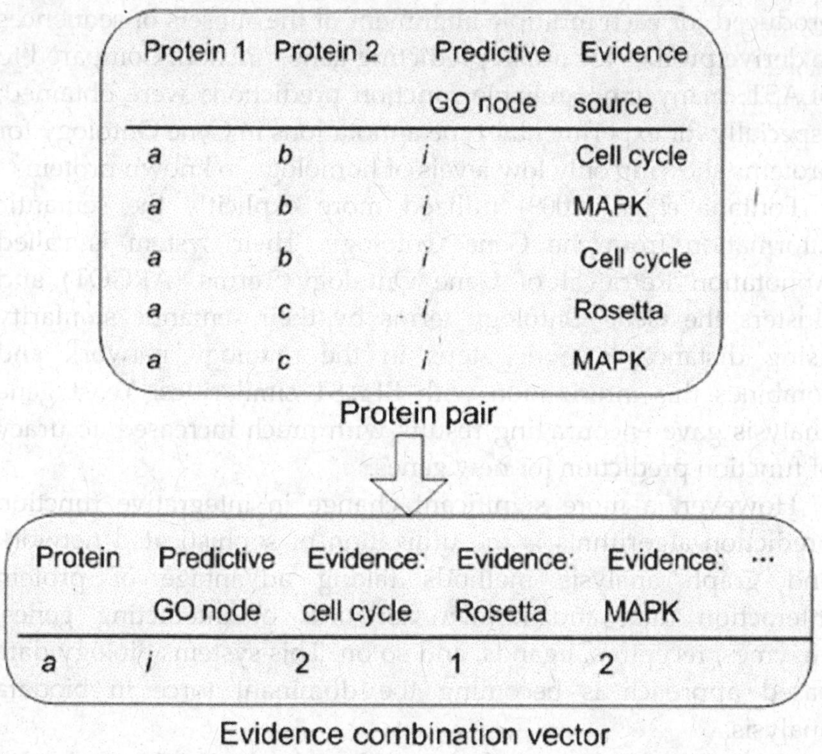

Protein 1	Protein 2	Predictive GO node	Evidence source
a	b	i	Cell cycle
a	b	i	MAPK
a	b	i	Cell cycle
a	c	i	Rosetta
a	c	i	MAPK

Protein pair

Protein	Predictive GO node	Evidence: cell cycle	Evidence: Rosetta	Evidence: MAPK	⋯
a	i	2	1	2	

Evidence combination vector

Figure 6-4. Construction of the feature vector for the Global Mapping of Unknown Proteins (GMUP) system of Xiong *et al.* (2006). Reproduced by Creative Commons Attribution License from an Open Access article.

effective in predicting functions for genes associated with the ribonucleoprotein complex which are important in protein translation from RNA, protein metabolism and protein transport. Several new candidate genes for specific functions were also found, serving as indicators for further experimental validation.

Recently, further improved and more sophisticated function annotation systems have appeared. The ConFunc (Wass & Sternberg 2008) system clusters sets of sequences identified by PSI-BLAST into multiple alignments based on their Gene Ontology annotations. A position specific scoring matrix is produced for each multiple alignment of the subsets of sequences to derive profiles for use in predicting gene function. Compared to BLAST, many more reliable function predictions were obtained, especially for experimental-type annotations in Gene Ontology for proteins showing only low levels of homology to known proteins.

Fontana *et al.* (2009) utilized more explicitly the semantic information from the Gene Ontology. Their system is called Annotation Retrieval of Gene Ontology Terms (ARGOT) and clusters the Gene Ontology terms by their semantic similarity using distance in node steps in the ontology network and combines this information with BLAST similarities. Yeast gene analysis gave encouraging results with much increased accuracy of function prediction for new genes.

However, a more significant change in integrative function prediction algorithms is the utilization of sophisticated network and graph analysis methods taking advantage of protein interaction data and known networks of interacting genes, enzymes, receptors, ligands, and so on. This systems biology data based approach is becoming the dominant force in biodata analysis.

The necessary large-scale input data for such systems is also becoming available. For example, the STRING database contains data on interactions of over 2.5 million proteins from 630 different organisms (Jensen *et al.* 2009). The principle of the integrative systems biology based approach is to find first sets of genes/proteins which are known to interact or which are correlated in expression level in some way. Then one looks for known network connections between these genes/proteins to annotate the nearby genes/proteins with the same function. The essential idea is to annotate linked unknown proteins based on distance from linked known proteins. A simplified example is shown in Figure 6-5.

A good general review of algorithms in this recent field of network-based function prediction is in Sharan *et al.* (2007), and the latest developments are discussed in Nabieva & Singh (2009).

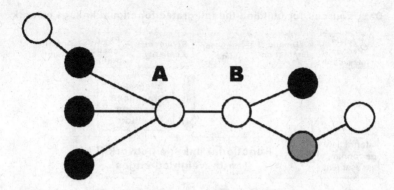

Figure 6-5. Examples of annotating linked proteins from network graph data. Proteins with two different known functions are in black and grey, unknown in white. The proteins in black having the same known function suggest the annotation of the linked white nodes in the network with the same "black" function. Note that protein marked A has the stronger evidence weight for "black" function (three one-step links to left and one two-step links to right, *e.g.* simple weighting 3x1 + 1x½ = 3½) than protein B (three two-step links and one one-step link, i.e. 3x½ + 1x1 = 2½). B may be bifunctional, but more likely has "black" function rather than "grey" function.

Many different approaches for utilizing the linkage information from the graph have been developed in the general data integration field, foremost among them Markov clustering (MCL, Enright *et al.* 2002) and Markov random fields (Deng *et al.* 2003) and various other graph clustering methods. A full treatment of this topic is not possible here, as the width of the field could justify an entire book by itself. Just some examples are given to illustrate recent innovative and interesting methods that are possible. In this young research field there is no consensus yet on which methods are most suitable for various types of bioinformatics data and predictions.

As an aside, a methodological note: for any integrated method, all available information should be used to aid the training of the function predictor. This is borne out in the study of Zhao *et al.* (2008), who developed an algorithm which utilizes also negative known samples using two-class SVMs to identify representative negative examples from unknown data, reducing the number of

Data sources for building the integrated functional linkage network

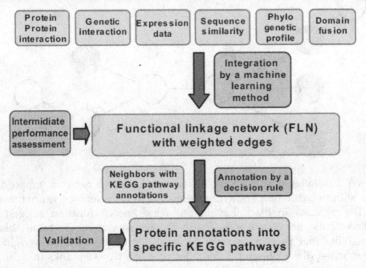

Figure 6-6. Function annotation network generation using an integrated approach. Multiple data sources are analysed first by a machine learning technique (*e.g.* linear SVM), followed by decision rules (*e.g.* Maximum Weight) for function annotations, which can be validated using KEGG pathways database. From Linghu *et al.* (2008). Reproduced by Creative Commons Attribution License from an Open Access article.

putative candidates to be considered and thus improving function prediction. Similarily, Persikov *et al.* (2008) used SVM to predict transcription factor binding, and found that including specific data on which amino acid residues do NOT bind the target DNA improved prediction accuracy markedly. Thus negative data can be very useful in enhancing classifier performance.

A well-performing system called FLN (Functional Linkage Network) was recently built by Linghu *et al.* (2008), and is depicted in Figure 6-6. Multiple types of data are used to generate putative linkages between proteins and then the obtained network is annotated by decision rules based on proteins having KEGG pathway annotations. KEGG pathway annotation data itself can be used for validation using withheld known data as an independent training set. Standard linear SVM was shown to work well in this system.

A good general review of recent integrative methods for protein function prediction using machine intelligence is that of Zhao *et al.* (2008b), who list a large number of techniques or unsupervised, semi-supervised and supervised classification. They also developed their own method of function annotation, which relies on calculating the shortest paths from known proteins to the protein to be annotated (Zhao *et al.* 2008c). The best weighting of such evidence is still an active research topic, as well as evaluating the effect of the topology of the network on weighting. Some types of network connections may also be "stronger" than others, *e.g.* direct receptor-ligand physical interaction should be considered weightier evidence than a protein binding to a promoter of another protein, where the effect in the cell is likely to take more time and be affected by other confounding factors. The time-scale of the interaction is thus also important to consider.

The next level of further integrated analysis could be simultaneous prediction of protein structure, function and interactions using multidomain data at the same time to find multidomain compatible predictions that best fit the data. Tress *et al.* (2008) and Rentzsch & Orengoa (2009) review the power of integrating data and the necessary tools that could be integrated in such next generation tools in the coming years.

6.6 Text mining bioliterature for automated annotation

Text mining is finally starting to arrive to bioinformatics and represents the very cutting edge of data mining techniques demanding linguistics, semantics, ontologies and grammar-based parsing methodologies, which are not so familiar to many bioinformaticians. There is already a huge amount of biological scientific literature, for example, MedLine contains over 18 million abstracts of articles, and an increasing proportion of scientific publications is coming available as full digital text, following the trend to open source publishing.

However, there are still many generic problems facing efficient biological text mining, in common with digital libraries (see Hull *et al.* 2008): metadata about publications is not standardized nor often attached to the publication itself, location of the data is not standardized, nor the publication identification codes, and authors are not uniquely identified either. Citations are also often not adequate to follow the citation networks properly to find related publications that could give additional related information. Additionally, the inherent problem of the texts themselves is the inconsistent nomenclature and grammar and writing style in articles by different authors and in different journals.

Text mining methods can be grouped into the following main categories: statistical and rule based methods, co-occurrence analysis and semantic analysis by natural language processing and miscellaneous other methods. A short introduction to this field for bioinformaticians is in Cohen *et al.* 2008. Below illustrative examples of modern natural language processing (NLP) and semantic methods are discussed, including semantic profiling and latent semantic analysis.

6.6.1 *Natural language processing (NLP)*

In NLP, the text is parsed by grammar analysis and semantic relations extracted by rules, verbs denoting assignments of functions, like "A inhibits B". Modern NLP systems are already quite good in processing standard English text, *e.g.* newspaper articles and general topic texts. However, processing of advanced scientific texts is more difficult and current NLP systems still need significant customization and nomenclature development for this kind of tasks.

At the current state of technology, it seems appropriate that text mining is initially used as an assisting tool for manual bioinformatics annotation. A recent study showed that NLP for extracting relevant annotation data from literature speeded up manual biodatabase curation at most 30%, which is encouraging, but still far from the ultimate goal of automated annotation (Alex *et al.* 2008). The BioCreative Consortium (www.biocreative.org) is evaluating automated text mining based annotation of protein interactions from full text scientific articles, and is progressing well to proof of concept demonstrations, but much more needs to be done to utilize the full potential of the information available in text form (Krallinger *et al.* 2008).

Regulatory networks of gene expression can also be literature mined, as shown recently by Aerts *et al.* (2008), who first used a training set of about 3,600 PubMed abstracts related to transcription regulation and promoter binding sites to evaluate all 16 million PubMed abstracts by cosine vector method and SVM method. Comparison of the top ranked abstracts and PubMed identifiers from the TRANSFAC transcription factor database showed a significant enrichment of gene regulatory articles, and corresponding DNA sequences could be extracted from full text articles in 25% of cases. This is a proof of principle for automated literature mining for interactions between transcription factors and transcribed genes. The system can effectively prioritize scientific articles for manual curation and for hyperlinking to accelerate genome and gene regulatory network annotation.

The literature mining method is getting powerful enough for even establishing new disease-pathway connections and drug discovery targets for biotechnology companies, as shown recently by Li & Agarwal (2009). They used PubMed Medical Subject Headings (MeSH) to mine from literature over 4,000 disease associated genes, mapped over 2,000 pathways to 600 diseases and generated a disease network which revealed a large number of novel disease-pathway associations or novel disease pairs linked by a common pathway. Simple shortest path calculations for related portions of the network graphs and CytoScape visualization software sufficed for new knowledge generation, even if the literature data can be considered a quite noisy data source.

6.6.1.1 *Integrating text mining with other data*

Small-scale text mining efforts for bioinformatics applications have already been successful in some special applications, especially when text mining is combined with other types of data. An early example is the work of Natarajn *et al.* (2006), who first extracted a list of genes which microarray analysis indicated are related to cancer invasiveness. Then scientific article full texts from 20 journals for the period of 1999–2003 were mined for putatively interacting proteins and the data used to build a protein interaction network based on these interacting proteins. The network was consistent with biology experts' opinions and also indicated new interesting interactions suitable for experimental verification.

In an interesting reverse strategy to that of Natarajan *et al.* above, Li *et al.* (2006) developed the LMMA (Literature Mining — MicroArray) method. They first extracted over 23,000 PubMed abstracts related to angiogenesis and mined gene names from them to obtain a set of 1,929 angiogenesis genes. Co-occurrence of the gene names in the same abstract was considered to indicate possible interaction. An interaction network was built based on the literature retrieved co-occurrences of the gene names. Out of these 1,929 genes, a set of about 1,200 genes were found in two cancer angiogenesis related microarray datasets. This subset network extracted from the microarray based network was shown to be of high quality with few false positives. The genes most strongly connected in the network turned out to be known cancer related genes, such as VEGF and other angiogenesis related genes, as validated by comparison to KEGG pathway database. This clearly demonstrates the added value of text mining in improving the microarray derived coexpression based networks. Recall (percent of correct interaction predictions) is still low, though, only about 50%. Therefore it will obviously be advantageous to include other additional data sources as evidence for likely or unlikely interactions in the network.

An even more attractive method could be to pool the literature mining and microarray data (and possibly other data) to a merged dataset to analyse them simultaneously, so that maximum information could be utilized. Then the weighting of the evidence will become also important. Such an approach appears not to have been reported yet.

In text mining for protein annotation, various existing annotation databases can be utilized to enhance text mining. Good results were obtained by Jaeger *et al.* (2008) using Gene Ontology as a starting point via sequence similarity. They could validate 80% of Gene Ontology annotations for proteins which had highly similar orthologous genes, and annotation precision was 100%. Several other similar text mining efforts have recently been reported, *e.g.* for protein interaction network induced by hepatitis C virus infection (De Chassey *et al.* 2008), gene-disease associations (Ozgur *et al.* 2008) and gene list clustering based on literature mining (Huang *et al.* 2008b).

The role of curated interaction databases like BIND, MINT and Reactome are thus becoming more important to include in such text mining systems, as the information in such databases is of considerably higher quality and better structured than in PubMed abstracts or in full text scientific articles. Web based biomedical text mining systems are also becoming available, including ALI BABA, EBIMed and PolySearch, which mine MedLine abstracts. An even more advanced recent system is SciMiner (Hur *et al.* 2009), which accepts gene lists as input and subsequently mines MedLine abstracts and downloads for analysis any available full text scientific articles as well and uses also Gene Ontology nomenclature, MeSH terms, HUGO gene nomenclature and so on. It achieves an impressive 87% recall and 71% precision for correct identification of gene and protein names.

There is no doubt that integrated data mining in high throughput biology will in the near future involve text mining in one form or another, integrated with other high-throughput data analysis of multidomain data, including gene and protein interaction, shared sequence motif and pathway databases (Yang *et al.* 2009). Thus the worth of text mining hinges on the additional value given by the text mined information, compared with the other more structured data sources.

It is now time for the text mining community to show that the computational effort and complexity of natural language processing is worth the effort, compared to the semi-automated annotation coming from the new high throughput experimental technologies, like EST tag sequencing, microarray data, protein interaction screens, proteomics and metabolomics profiles and so on. It will likely take quite a while before the multifarious diverse datasets can be adapted to "talk to each other" so that semantic information is harmonized without too much information loss. This effort could be termed "babelomics", in which the natural language processing and text mining experts should have an important role to play. The semantic methods for metadata are therefore very important, and are discussed in the next subchapter.

6.6.2 *Semantic profiling*

In the Semantic Web language, the semantic relations can be coded in RDF (Resource Description Format) formulation as triples, Subject => Relationship => Object, for example to state Enzyme X => phosphorylates => Protein Y. Such semantic triples form the basis of pathway and protein interaction databases, which, however, currently use incompatible nomenclatures and ontologies at different generalities of meaning. Although many of the current bioinformatics tools do not yet conform to the grand vision of RDF standards (*e.g.* relying on mere XML formalism for data exchange), ultimately large biological databases are envisaged to become available in RDF triple form of unitary semantic relations as data points, as indeed is planned for the whole of the internet shared information infrastructure (Bizer *et al.* 2008).

Such RDF semanticization of biodata has already started in earnest, as shown by the project Bio2RDF (Belleau *et al.* 2008, website **www.bio2rdf.org**). This project aims to sift through major biological databases to extract and make available RDF/XML formatted biological concepts and relationships for automated data analysis and linking using Universal Resource Indicators (URIs). In April 2008 it already contained some 2.4 billion data triplets, and envisages linking all biological knowledge semantically, so that they can be searched using semantic query engines using the SPARQL semantic query language. The hoped result is a kind of semantic web atlas of postgenomic knowledge for easy use by both humans and software agents. Searching for biological data would then be based on semantic type queries, rather than rigid text strings of standardized gene names, for example. The data will then be a global warehouse of conceptualized information available in its totality, with distributed storage over the internet.

Apart from these ambitious plans, which will take a while to be adopted as a goal by bioinformaticians and accepted by biologists, there have been more limited, but quite successful efforts to utilize semantic formalism in biodata mining in narrower data domains. A few examples are discussed below. General introductory reviews in this new field include Kim *et al.* (2008) who developed the GENIA corpus of biomedical concepts, parts of speech, syntactic trees and standard term lists. Dai *et al.* (2008) survey recent systems, including 13 tools for Named Entity Recognition (including GENIA tagger), 4 biodata semantic concept dictionaries (corpora), 2 biological event corpora and 12 web based services for biodata text mining. For example, iHOP retrieves sentences with the specified terms for genes and proteins and outputs graphs showing significant links between related genes and proteins and links to pathways and interaction databases (Fernandez *et al.* 2007).

Two main current approaches for handling semantic data are semantic profiling (discussed in this subchapter) and matrix factorization based methods (including latent semantic indexing and non-negative matrix factorization, discussed in the next subchapter). In semantic profiling, clusters of semantically related data items are created, and used for training a classification algorithm, which finds out in which semantic cluster(s) the new data items most likely belong to. Thus ontological concept distance between data points can be used as a classification variable, just as any other variable.

For example, semantic distance within the Gene Ontology classification tree between microarray genes can be used as an aid to classifying microarray data (Ovaska *et al.* 2008). They developed a fast algorithm in R language to calculate efficiently various semantic similarity metrics. Then a two-dimensional hierarchical clustering can be made both by semantic similarity of gene labels and by gene expression patterns. Figure 6-7 shows such a two-dimensional clustering. The latest Gene Name Normalization (GENO) system verifies and links gene names to correct nomenclature with an impressive accuracy of over 85% (Wermter *et al.* 2009).

Semantic similarity can be based on either information content of the semantic label set (more often occurring semantic labels contain less information), or by graph distance measures between data points. Many graph distance measures are available, and it is not yet clear which are best for which type of data. One approach is to find the most informative common node between the data points (or first common ancestor in a hierarchical semantic classification), and calculate a semantic similarity measure (*e.g.* Czekanowski-Dice, Kappa, Resnik *etc.* measures, see Ovaska *et al.* 2008 for details and references).

6.6.3 *Matrix factorization methods*

The second major technique for using semantic label data is matrix factorization, which is mostly used for feature reduction and finding the relevant features from a large number of possible

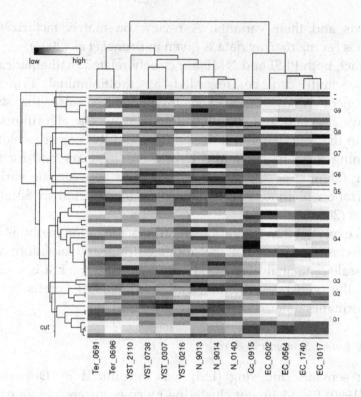

Figure 6-7. Combining semantic and gene expression clustering. Vertical dimension: Gene Ontology semantic label clustering, horizontal dimension: microarray gene expression pattern clustering. Grayscale indicates gene expression level. From Ovaska *et al.* (2008). Reproduced by BioMed Central Open Access Charter.

explaining variables in large text document databases (Shahnaz *et al.* 2006). The matrix of co-occurrence data of semantic labels can be processed by matrix factorization methods, in which the original data matrix of *e.g.* gene functions vs. genes can be partitioned to other matrices for cluster analysis and/or dimensionality reduction. The traditional method in text mining has been Latent Semantic Indexing (LSI, also called LSA for Latent Semantic Analysis) and its later probabilistic variant PLSI.

Recently, the method of Non-negative Matrix Factorization (NMF) was developed and used for bioinformatics data. Li & Ding (2006) give a lucid comparison of the NMF and PLSI methods and their variants. A review on matrix factorization methods for microarray data is given in Brunet *et al.* (2004).

In fact, both PLSI and NMF are closely related mathematically, because both can be regarded as multinominal Principal Component Analysis (PCA) methods, which solve the same objective function in a slightly different way. The algorithms do differ in the way they reach the solution and this can be exploited by joining them together in a hybrid ensemble method for text mining (Ding *et al.* 2006). The relationships between the various factorization methods are explained in more detail in Singh & Gordon (2008).

It is expected that the various matrix factorization methods and their hybrid variants will be adopted for many bioinformatics large-scale data analysis tasks. In the following the LSI is used to show the general principle and recent NMF methods for bioinformatics data are also briefly discussed.

6.6.3.1 *Latent Semantic Indexing*

Latent semantic indexing (LSI) was introduced by Deerwester *et al.* (1990) for document clustering by co-occurrence of words. It has been widely used in text data analysis, and is starting to be used in bioinformatics data analysis as well. The method first builds a word–document occurrence matrix X, which is then decomposed to three matrices (tri-factorization):

$$X = W \times S \times D$$

in which W represents the factors for words (rows), S is a diagonal matrix, which only the diagonal has non-zero values, and D is the matrix representing the factors for documents (columns). The decomposition is done so that when the three output matrices are multiplied, the original starting matrix X can be reconstructed.

The diagonal matrix values represent coefficients. A desired number of the largest coefficients can be selected and a smaller data matrix reconstructed by multiplication. In this way the dimensionality of the original data matrix can be reduced by orders of magnitude for easier analysis and without losing much information. Thus the information is compressed and less informative parts of the original matrix (words in the rows) are removed.

In bioinformatics applications of LSI, the words can be some characterizing features, *e.g.* sequence motifs, Gene Ontology labels, or any other biodata and documents can correspond to genes, proteins, genomes or drugs, depending on the purpose of the analysis. Figure 6-8 shows a simplified example of LSI decomposition for DNA motifs.

Singular value decomposition (SVD) has already been used a while in microarray analysis for informative ordering of genes to

```
DNA sequences:              Motifs:
D1: AAGCTTATGGGA            M1: AAGC
D2: CCGCCATCCCCC            M2: GGG
D3: TAAGCCGGGTCC            M3: AAAT
D4: AAATCTCTCCGC            M4: CC
D5: GTTGAAGCTCCT
D6: AATTTCCTCCTT
```

```
Occurrence matrix X:                   Motif factors W:
    D1 D2 D3 D4 D5 D6                   0.103  0.774 -0.002 -0.624
M1:  1  0  1  0  1  0         =         0.056  0.621  0.050  0.780      X
M2:  1  0  1  0  0  0                   0.044 -0.035  0.998 -0.039
M3:  0  0  0  1  0  0                   0.992 -0.114 -0.047  0.022
M4:  0  4  1  1  1  2
```

```
Diagonal matrix S:                            Sequence factors D:
4.828 0.000 0.000 0.000 0.000 0.000           0.033  0.822  0.239  0.215  0.227  0.411
0.000 2.076 0.000 0.000 0.000 0.000           0.672 -0.220  0.617 -0.072  0.318 -0.110
0.000 0.000 0.976 0.000 0.000 0.000    X      0.049 -0.194  0.000  0.974 -0.051 -0.097
0.000 0.000 0.000 0.655 0.000 0.000           0.238  0.137  0.272 -0.026 -0.919  0.069
                                             -0.698 -0.150  0.698  0.000  0.000 -0.050
                                              0.016 -0.444 -0.016  0.000  0.000  0.896
```

Figure 6-8. A fictitious example for the construction of word occurrence matrix and Singular Value Decomposition for latent semantic indexing from DNA sequences and DNA motifs in them. The occurrence matrix shows *inter alia* that the DNA motif M4 (CC) occurs 4 times in sequence D2. The occurrence matrix X is then decomposed to three multiplicative components W, S and D. Multiplying these three factors regenerates the original data matrix X. In the diagonal matrix S the underlined value 0.655 contains the lowest value on the diagonal, and this row could be deleted to reduce the dimensionality of the data by reconstructing a smaller matrix by multiplication.

co-varying groups (Liu *et al.* 2003). LSI tri-factorization was also used by Fukushima *et al.* (2008) for reducing the dimension of dimensions of over 2,300 microarray datasets for easier analysis of *Arabidopsis* plant gene expression patterns and correlations. They found that 20–40 factors normally sufficed to reconstruct a smaller version of the original data matrix without losing essential classificatory information to infer the function of selected genes by Gene Ontology classification.

LSI has also been used for protein secondary structure prediction (Ganapathiraju *et al.* 2005) and more significantly, for remote protein homolog finding. Dong *et al.* (2007) used LSI to automatically extract short informative oligopeptide words from protein sequences as features and used the LSI condensed dataset as input to SVM. The system outperformed pure SVM methods and PSI-BLAST and looks a promising approach, especially because computational efficiency was only slightly worse than PSI-BLAST.

In another significant development Klie *et al.* (2008) analysed large proteomics datasets from Human Proteome Organization HUPO's Plasma Proteome Project, containing protein peptide data from MALDI mass spectroscopy and other experiments. For protein identification from peptide fragments, about 75 features appear to suffice to reconstruct the original data matrices and are a powerful way to reduce the data dimensionality for subsequent analysis. Thus LSI seems to have great potential for analyzing large complex datasets with many possible classifying features.

In other fields of bioinformatics LSI has not yet become widely adopted, due to its newness, and because many other more established methods with good statistical tools have been available. One barrier is the large amount of computation needed for the matrix operations. Also, the more recent probabilistic Latent Semantic Indexing (PLSI) and various Non-negative Matrix Factorization (NMF) methods are now also competing for attention, and are therefore discussed next.

Latent Semantic Indexing also has some disadvantages, for example, it assumes Gaussian distribution of data in rows and columns, while many datasets are Poisson distributed, as is typical of sparse data from real life situations. Also, the factorized matrices contain negative values, which are not easily interpreted. Probabilistic Latent Semantic Indexing (PLSI) was developed to address these shortcomings and to add a statistical dimension (probabilities of belonging to a certain class).

In Probabilistic Latent Semantic Indexing (PLSI) the joint occurrence probability is usefully factorized into two components (di-factorization, for details, see Li & Ding 2006) and only positive values are allowed in the matrices. This method has been shown to outperform LSI in text mining, and has been recently adopted for some bioinformatics feature reduction tasks. Chang *et al.* (2008) used PLSI for capturing essential gapped dipeptide motifs prior to predicting protein subcellular localization by SVM and obtained competitive results, bettering *e.g.* pure PSORT algorithm. Several extracted dipeptide locations also corresponded to the known locations of cellular targeting peptide motifs. The algorithm also performed well for proteins showing low homology to known proteins in the training set, reaching an impressive overall accuracy of 86%.

6.6.3.2 *Non-negative Matrix Factorization*

Another recent matrix di-factorization method is Non-negative Matrix Factorization (NMF), which is related to PLSI, as remarked previously. It differs from the previously discussed factorization methods by not allowing negative values in the matrices, simplifying the matrix operations and their interpretability. These methods can be considered attractive alternatives to LSI and PLSI. However, there are also tri-factorization variants of NMF (Yoo & Choi 2009) and development and comparison of these methods is an active research topic. NMF avoids problems in distance measures due to absence of negative values in the matrices. One disadvantage of NMF over LSI is though, that convergence and stability of the matrix is not guaranteed.

Only a few bioinformatics data have so far been analysed by NMF methods. These include biological text analysis and microarray gene expression data analysis. Biomedical texts were analysed with NMF method by Kim *et al.* (2007), showing that previously unrecognized gene relationships could be uncovered based on PubMed abstract words as classifiers. Similarly, Heinrich *et al.* (2008) used titles and abstracts of Pubmed to cluster similar sets of genes, but evaluation of the results is difficult in the absence of a proper validation dataset.

A recent general review of NMF methods worth reading is Devarajan (2008). Liu *et al.* (2008) and Liu & Yuan (2008) discuss NMF for microarray data analysis applications. Protein interaction data can also be processed by NMF [Greene (2008) and Hutchins *et al.* (2008) developed a NMF method for finding position-specific motifs in large sequence sets].

The NMF methods appear to perform comparably with LSI in microarray analysis and several papers have been published for this kind of application. Most recently Fogel *et al.* (2008) analysed successfully the well-known AML leukemia dataset and obtained excellent results, with only one AML case misclassified. The NMF methods certainly warrant further scrutiny for unsupervised clustering and feature reduction in microarray data. Other large-scale and high-dimensionality data should also be investigated by NMF.

References

Aerts, S., Haeussler, M., van Vooren, S., Griffith, O.L., Hulpiau, P., Jones, S.J., Montgomery, S.B., Bergman, C.M. & Open Regulatory Annotation Consortium 2008. Text-mining assisted regulatory annotation. *Genome Biol.* 9(2): R31.

Alex, B., Grover, C., Haddow, B., Kabadjov, M., Klein, E., Matthews, M., Roebuck, S., Tobin, R. & Wang, X. 2008. Assisted curation: does text mining really help? *Pac Symp Biocomput. 2008: 556–567.*

Alexa, A., Rahnenfuhrer, J. & Lengauer, T. 2006. Improved scoring of functional groups from gene expression data by decorrelating GO graph structure. *Bioinformatics 22: 1600–1607.*

Antonov, A.V., Dietmann, S., Wong, P., Lutter, D. & Mewes, H.W. 2009. GeneSet2miRNA: finding the signature of cooperative miRNA activities in the gene lists. *Nucleic Acids Research pp. 1–6 Advance Access published May 6, 2009.*

Antonov, A.V., Schmidt, T., Wang, Y. & Mewes, H.W. 2008. ProfCom: a web tool for profiling the complex functionality of gene groups identified from high-throughput data. *Nucleic Acids Res. 36: W347–W351.*

Belleau, F., Nolina, M-A., Tourigny, N., Rigault, P. & Morissette, J. 2008. Bio2RDF: Towards a mashup to build bioinformatics knowledge systems. *Journal of Biomedical Informatics 41(5): 706–716.*

Bizer, C., Heath, T., Idehen, K. & Berners-Lee, T. 2008. Linked Data on the Web. pp. 1–2 in: *Proceedings of Workshop at the 17th International World Wide Web Conference, Beijing, China, April 22, 2008 (LDOW2008).*

Brunet, J.P. *et al.* 2004. Metagenes and molecular pattern discovery using matrix factorization. *Proc. Natl Acad. Sci. USA 101: 4164–4169.*

Carbon, S., Ireland, A., Mungall, C.J., Shu, S., Marshall, B., Lewis, S., AmiGO Hub & Web Presence Working Group. 2009. Web AmiGO: online access to ontology and annotation data. *Bioinformatics 25(2): 288–289.*

Casadio, R., Martelli, P.L., Pierleoni, A. 2008. The prediction of protein subcellular localization from sequence: a shortcut to functional genome annotation. *Briefings in Functional Genomics and Proteomics 7(1): 63–73.*

Chagoyen, M., Carazo, J.M., Pascual-Montano, A. 2008. Assessment of protein set coherence using functional annotations. *BMC Bioinformatics 9: 444.*

Chang, J.M., Su, E.C., Lo, A., Chiu, H.S., Sung, T.Y. & Hsu, W.L. 2008. PSLDoc: Protein subcellular localization prediction based on gapped-dipeptides and probabilistic latent semantic analysis. *Proteins 72(2): 693–710.*

Chitale, M., Hawkins, T., Park, C. & Kihara, D. 2009. ESG: Extended similarity group method for automated protein function prediction. *Bioinformatics (in press).*

Chou, K.C. & Shen, H.B. 2008. Cell-PLoc: a package of Web servers for predicting subcellular localization of proteins in various organisms. Nature Protocols 3(2): 153–162.

Cohen, K.B. & Hunter, L. 2008. Getting started in text mining. *PLoS Comput. Biol. 4, e20.*

Dai, H-J., Lin, J. Y-W., Huang, C-H., Chou, P-H., Tsai, R.T-H. & Hsu, P-H. 2008. A Survey of State of the Art Biomedical Text Mining Techniques for Semantic Analysis. pp. 410–417 in: *Proceedings, Sensor Networks, Ubiquitous and Trustworthy Computing SUTC '08, IEEE International Conference.*

De Chassey, B. *et al.* 2008. Hepatitis C virus infection protein network. *Mol. Syst. Biol. 4: 230.*

Deerwester, S., Dumais, S.T., Furnas, G.W., Landauer, T.K. & Harshman, R. 1990. Indexing by latent semantic analysis. *J Am Soc Inf Sci 41: 391–407.*

Devarajan, K. 2008. Nonnegative Matrix Factorization: An Analytical and Interpretive Tool in Computational Biology. *PLoS Comput Biol. 4(7): e1000029.*

Ding, C., Li, T. & Pang, W. 2006. Nonnegative Matrix Factorization and Probabilistic Latent Semantic Indexing: Equivalence, Chi-square Statistic, and a Hybrid Method. *Pp. 342–347 in: Proc. of AAAI National Conf. on Artificial Intelligence (AAAI-06), July 2006.*

Dong, Q.W., Wang, X.L. & Lin, L. 2007. Application of latent semantic analysis to protein remote homology detection. *Bioinformatics. 22(3): 285–290.*

Eden, E., Narvon, R., Seinfeld, I., Lipson, D., Yakima, Z. 2009. GOrilla: a tool for discovery and visualization of enriched GO terms in ranked gene lists. *BMC Bioinformatics 10: 48.*

Fernandez, J.M., Hoffmann, R. & Valencia, A. 2007. iHOP web services. *Nucl. Acids Res. 35(Web Server issue): W21–6.*

Fogel, P., Young, S.S., Hawkins, D.M., Ledirac, N. 2007. Inferential, robust non-negative matrix factorization analysis of microarray data. *Bioinformatics 23(1): 44–49.*

Fontana, P., Cestaro, A., Velasco, R., Formentin, E. & Toppo, S. 2009 Rapid Annotation of Anonymous Sequences from Genome Projects Using Semantic Similarities and a Weighting Scheme in Gene Ontology. *PLoS ONE 4(2): e4619.*

Fukushima, A., Wada, M., Kanaya, S. & Arita, M. 2008. SVD-based anatomy of gene expressions for correlation analysis in Arabidopsis thaliana. *DNA Res. 15(6): 367–374.*

Ganapathiraju, M. *et al.* 2004. Characterization of protein secondary structure, Application of latent semantic analysis using different vocabularies. *IEEE Signal Processing Magazine 21: 78–87.*

Grant, J.R. & Stothard, P. 2008. The CGView Server: a comparative genomics tool for circular genomes. *Nucleic Acids Res. 36(Web Server issue): W181–184.*

Greene, D., Cagney, G., Krogan, N. & Cunningham, P. 2008. Ensemble non-negative matrix factorization methods for clustering protein-protein interactions. *Bioinformatics 24(15): 1722–1728.*

Gschloessl, B., Guermeur, Y. & Cock, J.M. 2008. HECTAR: a method to predict subcellular targeting in heterokonts. *BMC Bioinformatics 9: 393.*

Hawkins, T., Chitale, M., Luban, S. & Kihara, D. 2009. PFP: Automated prediction of gene ontology functional annotations with confidence scores using protein sequence data. *Proteins: Structure, Function, and Bioinformatics 74(3): 566–582.*

Heinrich, K.E., Berry, M.W. & Homayouni, R. 2008. Gene tree labeling using nonnegative matrix factorization on biomedical literature. *Comput Intell Neurosci. 2008: 276535.*

Horton, P., Park, K.J., Obayashi, T., *et al.* 2007. WoLF PSORT: protein localization predictor. *Nucleic Acids Res 35: W585–587.*

Huang, D.W., Sherman, B.T., Lempicki, R.A. 2008a. Bioinformatics enrichment tools: paths toward the comprehensive functional analysis of large gene lists. *Nucleic Acids Res 37(1): 1–13.*

Huang, Z.X. *et al.* (2008b) GenCLiP: a software program for clustering gene lists by literature profiling and constructing gene co-occurrence networks related to custom keywords. *BMC Bioinform. 9, 308.*

Hull, D., Pettifer, S.R., Kell, D.B. 2008. Defrosting the digital library: bibliographic tools for the next generation web. *PLoS Comput Biol. 4(10): e1000204.*

Hur, J., Schuyler, A.D., States, D.J. & Feldman, E.L. 2009. SciMiner: web-based literature mining tool for target identification and functional enrichment analysis. *Bioinformatics. 25(6): 838–840.*

Hutchins, L.N., Murphy, S.M., Singh, P. & Graber, J.H. 2008. Position-dependent motif characterization using non-negative matrix factorization. *Bioinformatics 24(23): 2684–2690.*

Jaeger, S., Gaudan, S., Leser, U., Rebholz-Schuhmann, D. 2008. Integrating protein-protein interactions and text mining for protein function prediction. BMC *Bioinformatics. 9 Suppl 8: S2.*

Jensen, L.J., Kuhn, M., Stark, M., Chaffron, S., Creevey, C., Muller, J., Doerks, T., Julien, P., Roth, A., Simonovic, M., Bork, P. & von Mering, C. 2009. STRING 8--a global view on proteins and their functional interactions in 630 organisms. *Nucleic Acids Res. 37(Database issue): D412–416.*

Kim, H., Park, H. & Drake, B.L. 2007. Extracting unrecognized gene relationships from the biomedical literature via matrix factorizations. *BMC Bioinformatics 8 Suppl 9: S6.*

Kim, J.D., Ohta, T., & Tsujii, J. 2008. Corpus annotation for mining biomedical events from literature. *BMC Bioinformatics 9: 10.*

Klie, S., Martens, L., Vizcaíno, J.A., Côté, R., Jones, P., Apweiler, R., Hinneburg, A. & Hermjakob, H. 2008. Analyzing large-scale proteomics projects with latent semantic indexing. *J Proteome Res 7(1): 182–191.*

Krallinger, M., Valencia, A., Hirschman, L. 2008. Linking genes to literature: text mining, information extraction, and retrieval applications for biology. *Genome Biol. 9 Suppl 2: S8.*

Li, S., Wu, L. & Zhang, Z. 2006. Constructing biological networks through combined literature mining and microarray analysis: a LMMA approach. *Bioinformatics 22: 2143–2150.*

Li, T. & Ding, C. 2006. The Relationships Among Various Nonnegative Matrix Factorization Methods for Clustering. *pp. 362–371 in: Sixth International Conference on Data Mining, 2006. ICDM '06.*

Li, Y. & Agarwal, P. 2009. A pathway-based view of human diseases and disease relationships. *PLoS ONE. 4(2): e4346.*

Lin, F.P., Coiera, E., Lan, R. & Sintchenko, V. 2009. In silico prioritisation of candidate genes for prokaryotic gene function discovery: an application of phylogenetic profiles. *BMC Bioinformatics 10: 86.*

Linghu, B., Snitkin, E.S., Holloway, D.T., Gustafson, A.M., Xia, Y. & DeLisi, C. 2008. High-precision high-coverage functional inference from integrated data sources. *BMC Bioinformatics 9: 119.*

Liu, L., Hawkins, D.M., Ghosh, S. & Young, S.S. 2003. Robust singular value decomposition analysis of microarray data. *Proc. Natl Acad. Sci. USA 100:* 13167–13172.

Liu, W. & Yuan, K. 2008. Sparse p-norm Nonnegative Matrix Factorization for clustering gene expression data. *International Journal of Data Mining and Bioinformatics 2(3): 236–249.*

Liu, W., Yuan, K. & Yea, D. 2008. Reducing microarray data via nonnegative matrix factorization for visualization and clustering analysis. *Journal of Biomedical Informatics 41(4): 602–606.*

Maltsev, N., Glass, E., Sulakhe, D., Rodriguez, A., Syed, M.H., Bompada, T., Zhang, Y., D'Souza, M. 2006. PUMA2–grid-based high-throughput analysis of genomes and metabolic pathways. *Nucleic Acids Res 34: D369–D372.*

Médigue, C., Moszer, I. 2007. Annotation, comparison and databases for hundreds of bacterial genomes. *Res Microbiol 158(10): 724–736.*

Meyer, F., Paarmann, D., D'Souza, M., Olson, R., Glass, E.M., Kubal, M., Paczian, T., Rodriguez, A., Stevens, R., Wilke, A., Wilkening, J., Edwards, R.A. 2008. The metagenomics RAST server - a public resource for the automatic phylogenetic and functional analysis of metagenomes. *BMC Bioinformatics. 9: 386.*

Mistry, M. & Pavlidis, P. 2008. Gene Ontology term overlap as a measure of gene functional similarity. *BMC Bioinformatics 9: 327.*

Nabieva, E. & Singh, M. 2008. Protein Function Prediction via Analysis of Interactomes. *Pp. 231–258 in: Janusz M. Bujnicki (Ed.) Prediction of Protein Structures, Functions, and Interactions. Wiley.*

Nair, R. & Rost, B. 2005. Mimicking cellular sorting improves prediction of subcellular localization. *J Mol Biol 348: 85–100.*

Natarajan, J. *et al.* 2006. Text mining of full-text journal articles combined with gene expression analysis reveals a relationship between sphingosine-1-phosphate and invasiveness of a glioblastoma cell line. *BMC Bioinform. 7: 373.*

Ovaska, K., Laakso, M & Hautaniemi, S. 2008. Fast Gene Ontology based clustering for microarray experiments. *BioData Mining 1: 11.*

Ozgur, A. *et al.* 2008. Identifying gene-disease associations using centrality on a literature mined gene-interaction network. *Bioinformatics 24: i277–i285.*

Persikov, A.V., Osada, R. & Singh, M. 2009. Predicting DNA recognition by Cys2His2 zinc finger proteins. *Bioinformatics 25(1): 22–29.*

Pierleoni, A., Martelli, P.L., Fariselli, P. *et al.* 2006. BaCelLo: a balanced subcellular localization predictor. *Bioinformatics 22: e408–416.*

Pierleoni, A., Martelli, P.L., Fariselli, P. & Casadio, R. 2007. eSLDB: eukaryotic subcellular localization database. *Nucleic Acids Res 35: D208–212.*

Reeves, G.A., Eilbeck, K., Magrane, M., O'Donovan, C., Montecchi-Palazzi, L., Harris, M.A., Orchard, S., Jimenez, R.C., Prlic, A., Hubbard, T.J., Hermjakob, H. & Thornton, J.M. 2008. The Protein Feature Ontology: a tool for the unification of protein feature annotations. *Bioinformatics 24(23): 2767–2772.*

Rentzsch, R. & Orengo, C.A. 2009. Protein function prediction – the power of multiplicity. *Trends in Biotechnology 27(4): 210–219.*

Shahnaz, F., Berry, M.W., Pauca, V.P. & Plemmons, R.J. 2006. Document clustering using nonnegative matrix factorization. *Information Processing & Management 42(2), 373–386.*

Sharan, R., Ulitsky, I. & Shamir, R. 2007. Network-based prediction of protein function. *Molecular Systems Biology 3: 88.*

Singh, A. S. & Gordon, G.J. 2008. A Unified View of Matrix Factorization Models. *Lecture Notes in Computer Science 5212/2008: 358–373.*

Sprenger, J., Fink, J.L. &Teasdale, R.D. 2006. Evaluation and comparison of mammalian subcellular localization prediction methods. *BMC Bioinformatics 7: S3.*

Stothard, P. & Wishart, D.S. 2006. Automated bacterial genome analysis and annotation. *Curr Opin Microbiol. 9(5): 505–510.*

Sun, C.-H., Kim, M-S., Han, Y. & Yi, G-S. 2009. COFECO: composite function annotation enriched by protein complex data. *Nucleic Acids Research pp. 1–6, Advance Access published May 8, 2009*

Tamura, T. & Akutsu, T. 2007. Subcellular location prediction of proteins using support vector machines with alignment of block sequences utilizing amino acid composition. *BMC Bioinformatics 8: 466.*

Tress, M., Bujnicki, J.M., Lopez, G. & Valencia, A. 2008. Integrating Prediction of Structure, Function, and Interactions. *Pp. 259–279 in: Janusz M. Bujnicki (Ed.) Prediction of Protein Structures, Functions, and Interactions. Wiley.*

Van Domselaar, G.H., Stothard, P., Shrivastava, S., Cruz, J.A., Guo, A., Dong, X., Lu, P., Szafron, D., Greiner, R. & Wishart, D.S. 2005. BASys: a web server for automated bacterial genome annotation. *Nucleic Acids Res 33: W455–W459.*

Wass, M.N. & Sternberg, M.J. 2008. ConFunc--functional annotation in the twilight zone. *Bioinformatics 24(6): 798–806.*

Wermter, J., Tomanek, K. & Hahn, U. 2009. High-performance gene name normalization with GeNo. *Bioinformatics 25(6): 815–821.*

Wu, X. & Watson, M. 2009. CORNA: testing gene lists for regulation by microRNAs. *Bioinformatics 25: 832–833.*

Xiao, F., Zuo, Z., Cai, G., Kang, S., Gao, X. & Li, T. 2008. miRecords: an integrated resource for microRNA-target interactions. *Nucleic Acids Res. 37(Database issue): D105–110.*

Xiong, J., Rayner, S., Luo, K., Li, Y. & Chen, S. 2006. Genome wide prediction of protein function via a generic knowledge discovery approach based on evidence integration. *BMC Bioinformatics 7: 268.*

Yang, Y., Adelstein, S.J. & Kassis, A.I. 2009. Target discovery from data mining approaches. *Drug Discov Today 14(3–4): 147–54.*

Yoo, J. & Choi, S. 2009. Probabilistic matrix tri-factorization. *pp. 1–4 in: Proceedings of the IEEE International Conference on Acoustics, Speech, and Signal Processing (ICASSP), Taipei, Taiwan, 2009 (in press).*

Yu, C., Zavaljevski, N., Desai, V., Johnson, S., Stevens, F.J., Reifman, J. 2008. The development of PIPA: an integrated and automated pipeline for genome-wide protein function annotation. *BMC Bioinformatics 9: 52.*

Zhao, X.M., Chen, L.N., Aihara, K. 2008a. Gene function prediction using labeled and unlabeled data. *BMC Bioinformatics 9: 57.*

Zhao, X-M., Chen, L & Aihara, K. 2008b. Protein function prediction with high-throughput data. *Amino Acids 35: 517–530.*

Zhao, X-M., Chen, L & Aihara, K. 2008c. Protein function prediction with the shortest path in functional linkage graph and boosting. *International Journal of Bioinformatics Research and Applications 4(4): 375–384.*

Zhu, H., Hu, G.Q., Yang, Y.F., Wang, J., She, Z.S. 2007. MED: a new non-supervised gene prediction algorithm for bacterial and archaeal genomes. *BMC Bioinformatics 8: 97.*

Chapter 7

New Methods for Genomics Data: SVM and Others

7.1 SVM kernels

Support Vector Machine is a generic successful method developed by Vapnik in the 1990s. The method essentially transforms the data to a multidimensional vector space in which different classes or clusters of data points can be separated by maximum margin hyperplanes. The simplest case, the linear SVM is a remarkably powerful classifier and is widely used in all data mining applications. Better fitting non-linear boundaries between groups of datapoints belonging to different classes can be obtained using polynomial or Gaussian kernels. Examples of various separating boundaries are shown in Figure 7-1. Note that making the boundary to fit too closely the groups to be separated will lead to overfitting and will degrade generalization performance (Figure 7-1, right). This should be kept in mind when optimizing the kernel parameters, so that both prediction accuracy and generalizability are balanced appropriately. This is generally best done by having several independent validation datasets.

The use of standard SVM techniques is already well established also in bioinformatics and good general books and reviews are available to get started. A good technical backgrounder is available in the book by Shawe-Taylor & Cristianini (2004) and required reading for bioinformatics applications of SVM is the book by Schölkopf *et al.* (2004). Ben-Hur *et al.* (2008) gives a short

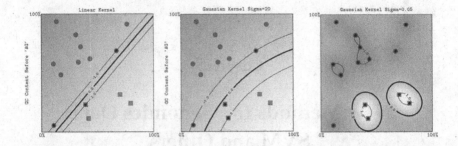

Figure 7-1. Examples of maximum margin boundaries between two groups of data points (circles and squares) generated by linear kernel (left) and Gaussian kernel (middle, right) in a support vector machine. Some parameter settings can result in overfitting (shown at right) and leads to poor generalizability and unclassifiable data points. Adapted from Ben-hur *et al.* (2008). Reproduced by Creative Commons Attribution License from an Open Access article.

accessible introduction of SVM using a bioinformatics example for quick understanding of the basic concepts. Here these basics will not be reiterated, but rather in the following recent developments in specific promising directions are discussed, including hybrid methods combining SVM with other techniques.

SVM is based on the power of kernels to process the data into a suitable form for efficient and sensitive large-scale classification. In essence the kernel trick allows use of a linear algorithm for a non-linear problem. It is remarkable, though, that often the simple linear kernel shows already very good classification performance for even complex problems.

A large variety of different kernels have been developed, tuned to the specific features of datasets to be analysed and to the types of features that are thought to contain the best classificatory features. Only some basic interesting kernels are discussed here to show the variety of possible methods, as well as new promising hybrid methods involving SVM.

SVM methodology has matured so much, so that focus is moving to large-scale applications and kernel learning, *i.e.* developing algorithms that choose and optimize the best kernel for the problem and dataset at hand, or combine different kernels together This trend is fascinating, as metalearning in general is gaining proponents, purported to aid scientists to choose objectively and automatically the method best suited to the problem. Bioinformaticians should also follow this work to adopt suitable metalearning methods. Metalearning approach has already been used for selecting the best clustering algorithms for gene expression data (de Souto *et al.* 2008a, 2008b, Souza & Carvalho 2008).

String kernels are naturally suited for DNA and protein sequence analysis. Many different string kernels have been suggested and used successfully for many bioinformatics problems and a full treatment of this wide field is not intended here. In the first major type of string kernels, sequence composition features are used directly to obtain a measure of similarity. In the simplest such string kernel one counts the number of common words between two sequences (bag of words kernel), the frequencies of n-grams or k-mers (spectrum kernel), or finds which k-mers are present in both sequences (weighted spectrum kernel). As we have seen elsewhere in this book, fast suffix tree methods are available to obtain the necessary information for the k-mer based string kernels, so that they have gained popularity due to their computational efficiency.

The second major type of string kernels is based on a generated model of a sequence (*e.g.* Hidden Markov Model) or scoring a local alignment. Such kernels include Fisher kernel and TOP kernel. Whatever the string kernel formalism, the main point is that it should be easily calculated or at least extract good classificatory features for the problem at hand. Only recently have efficient algorithms and faster personal computers made string kernels competitive in terms of necessary computing time, so that string kernels are expected to become very useful for many bioinformatics problems, which previously have been analysed by other methods, HMMs, for example. Sophisticated string kernels are already showing utility for *e.g.* gene regulatory motif finding (Schultheiss *et al.* 2009).

Large-scale data analysis is becoming increasingly important in the post-genomic era, and therefore the large-scale string kernel approach of Sonnenburg *et al.* (2007) deserves special mention. The sparse string kernels are based on sparse representation of the k-mer information, reducing the dimensionality of sequence comparison markedly. In addition, it is possible to get a weight for the contribution of each k-mer in the discrimination performance of the SVM analysis. For efficient use of sparse kernels, storage of the k-mer arrays is important and depends on the size of the sequences to be analysed. For the largest genome scale datasets suffix trees may be the best, but for medium size data sets sorted arrays may be most suitable.

Stem kernels are an extension of string kernels, in which the kernel searches for all RNA base-pairings in potential stem structures along the sequence. Secondary structure of RNAs depends on specific paired segments of the molecule which are energetically the most favourable. Computation of all possible pairing segments along the self-folding RNA soon becomes infeasible using exhaustive energy calculations of all possible folding structures. Even finding all possible common base pairs and matching segments (stems of the microRNAs) of any length using stem kernels (Sakakibara *et al.* 2007) scales to the fourth power of the length of the input sequence. Some speedup has been achieved by a recent formulation using directed acyclic graphs (Sato *et al.* 2008), but further efficiency gains are needed to make finding all possible folding structures in longer sequences or full genomes feasible. Therefore limiting the search space is necessary.

Such search space partitioning has recently been suggested by Morita *et al.* (2009), who developed a new base-pairing profile local alignment (BPLA) kernel enabling fast data generation for SVM analysis. In spite of its simplicity (local pairwise alignments only used as features), it appears to perform as well as the state-of-art miPred. The miPred method relies on a total of 29 different structural features in the training set in SVM analysis. In BPLA kernel SVM the local alignment reduces the search space so that the new method scales in linear time of kernel calculation and in $O(x3 + y3)$ time and in $O(x2 + y2)$ memory for the local base-pairing probability matrix of segments x and y being compared.

Morita *et al.* (2009) have already run their algorithm using a linux cluster of 20 CPUs for one week to predict all snoRNAs in the nematode *Caenorhabditis elegans* genome, which is about 97 megabases long. The full genome analysis discovered many novel non-coding RNA candidates (some of which were even validated experimentally) and outperformed the state of art RNAz method for known RNAs. This method should be evaluated for other genomes using a good benchmark and experimental validation. The remaining problem is fast optimization of the SVM hyperparameters for each dataset, but this should be overcome by new optimization techniques. This method would then be a big step toward fully automated annotation of genomes for non-coding RNA candidates, including the microRNAs. It remains to be seen, if further reduction in time and memory requirements could be made. In any case, this method would be a good candidate for GPU acceleration as well, as the kernel is parallelizable, so that a fast GPU implementation may already be sufficient for large-scale genome scans with this algorithm.

Graph kernels are designed to suit graph-based data, for example pathway, protein interaction network or Gene Ontology type data, but other types of data can also be manipulated to suit graph kernels, if the problem at hand can be formulated as a graph search space exploration. The book by Shawe-Taylor & Cristianini (2004) discusses the basic theory of graph kernels in detail; only topical recent advances and promising applications of graph kernels are discussed here.

Graph kernels aim to represent *e.g.* molecular structures or interaction networks in a form analysable by SVM. This approach has already been very successful in a number of domains, like social network analysis, physics and engineering. In effect, the graphs are transformed into n-dimensional vector spaces, and a general method of using this by using graph edit distance has been recently proposed (Bunke & Riesen 2008). This is conceptually similar to the concept of string edit distance used in sequence similarity algorithms for aligned DNA or protein sequences. A large variety of 2D and 3D structure graph kernels has already been developed for proteins and small molecules (see the review of Vert & Jacob 2008 for a classification and examples of them).

Graph kernels are a new promising approach also for comparing protein structures. The structure of the protein is coded in a kernel as a network representation of links and proximities of protein segments and amino acids. Graph kernels were earlier thought to be too compute-intensive, but recently accelerated computation formulations have been developed, paving the way for more extensive use of graph kernels in protein structure comparisons (Borgwardt *et al.* 2007) and even in high-level feature extraction from photographic images (Vert *et al.* 2009).

Graph kernels are naturally suited to biological network data analysis. One example is the diffusion kernel approach of Sun & Ye (2008). In a diffusion kernel, local information is diffused through linking nodes by repeated multiplications, controlled by a parameter that regulates the strength and extent of diffusion (*i.e.* how far in the network the links are extended to). In their diffusion kernel the diffusion parameter is tuned to optimum by learning the large margin criteria between the known classes to be separated. A good performance diffusion kernel could be obtained by comparing 60 different kernels with different diffusion parameter values, and the optimized kernel was then applied to two large protein interaction datasets with good results.

The graph kernels discussed above are based on an assumption of independence of the nodes from the rest of the network for statistical evaluation. They are supervised methods relying on some training set to guide finding the most likely network. However, a more realistic assumption is that the nodes are not independent, but often depend on at least nearby nodes. Thus unsupervised network building is needed for novel datasets or for an unbiased starting point in evaluating a network. Lippert *et al.* (2009) developed such a graph kernel formulation, in which unsupervised network building and non-independence of nodes is taken into account. This new method looks very interesting to apply to protein interaction networks and other biological network data, for which no training data is available.

However, it should also be remembered, that other graph matching algorithms can be utilized and can be very effective. For example, Zaslavskiy *et al.* (2009) used latest graph matching algorithms to solve a global network alignment problem for protein-protein interactions. This underscores the importance of exploring different types of algorithms for each problem set. For very large and heterogeneous complex networks the SVM graph kernels may not in fact be the best tool, though they should work well from smaller networks.

The **Tanimoto kernel** is a well established kernel for classical fingerprints of chemical descriptors (as discussed in chapter 4.12.) for comparing molecular structures. Tanimoto kernel is important in chemogenomics, where in addition different kernels are being combined, the Tanimoto kernel taking care of molecular structure similarity and other kernels being used for DNA or protein sequence similarity or interaction data. Recent methods in this area are summarized in Jacob & Vert (2007). They have used a Tanimoto kernel together with a hierarchical kernel for KEGG pathway data to obtain classification accuracies of 82–88% for enzyme, GPCR and ion channel function/target prediction, even though especially GPRs (G-protein coupled receptors) have been considered a very difficult group of proteins for target prediction (Jacob & Vert 2008). Their combined two-kernel system can predict even targets which have no known ligands. Local sequence alignment based kernels for targets did not work so well, which underscores the difficulty of identifying the relevant sequence-based features responsible for ligand-binding.

Multiple kernel learning is another important new direction, which aims to evaluate and optimize a sufficiently large number of different kernels to obtain a better solution and to understand the importance of different features better. This is related to the metalearning concept mentioned earlier. Multiple kernel learning is currently the leading method for video image analysis, but could be used for bitmap images of molecular structures, and for biological tissue images as well. New modifications of the method appear to be performing very well (Yang *et al.* 2008, Gönen *et al.* 2008) and thus are worth exploring for bioinformatics applications as well.

For large data sets and large numbers of kernels, Semi-Infinite Linear Program based multiple kernel learning is one possible direction, and has been used in prediction of human gene DNA transcription start sites (Sonnenburg *et al.* 2006) The recent techniques are reviewed in Kashima *et al.* (2009) and Kloft *et al.* (2008) developed a new non-sparse multiple kernel learning method. Most recently, Gehler *et al.* (2009) generalized multiple kernel learning (MKL) to infinite kernel learning (IKL) for image analysis. This method would be interesting to apply to bioinformatics datasets as well.

The recent development of a multitude of new kernels for many types of bioinformatics data exemplifies the strength of the kernel based SVM methods in general in computational biology. In general, the feature extraction and selection problem can often be solved successfully by recasting the problem toward good kernel design. Thus the future seems bright for competent kernel designers in large-scale bioinformatics projects, especially for development of string kernels and graph kernels as well as metalearning multiple kernels.

7.2 SVM trees

SVM is powerful in many ways, as shown by the hundreds of publications in the bioinformatics domain reporting use of SVM in biodata analysis, but for large datasets and search spaces, computational complexity is still often limiting for finding the best classification in reasonable time. Therefore ensembles of

SVMs have been explored to partition the problem to smaller manageable chunks. The partitions can be pre-determined, or made on the fly, iteratively, in which case the easily identified data points are separated by the first SVM, then the remainder by the next node in a branching tree fashion. Such a SVM-tree type classification method, called 2-SVMT, was developed by Pang *et al.* (2006) for cancer microarray analysis with class imbalanced datasets. Classification is obtained by recursive partitioning and each partition classified by an SVM node in a series of data partitions having certain marker genes relevant in the class prediction for each of these subspaces (see Figure 7-2).

Such successive iterative partitioning and SVM classification appears to be most useful for datasets which are biased in their class distribution (*i.e.* much more samples in class 1 than in class 2 or *vice versa*). The classification was also more robust than single SVM analysis, KNN or C4.5 methods, as shown by ROC analysis. SVM ensembles have been found in general to be useful to analyse data with a class imbalance (Yan *et al.* 2003). A useful feature of the 2-SVMT method is that the generated decision tree structure can give additional information about possible subclasses within the two classes, and even may indicate outliers or difficult to classify data items in the dataset.

The method can be extended to multiple classes (m-SVMT), in which case each decision node generate as many branches as there are classes. Thus a decision tree structure is constructed, and that tree structure could be utilized for making decision rules about classification. Pang & Kasabov (2008) developed such an algorithm to analyse cancer classification data with good results. The method combines the easy comprehensibility of decision rule methods with the accuracy of the SVM classification and could be used to analyse many other kinds of bioinformatics data as well.

Similar SVM tree hierarchical classification systems have been used to analyse medical databases (Chow *et al.* 2008) and other large databases (Fehr *et al.* 2008), medical video images (Gao *et al.* 2009), and microarray gene data for gene selection (Mundra & Rajapakse 2007). Such constructed decision trees may reveal new

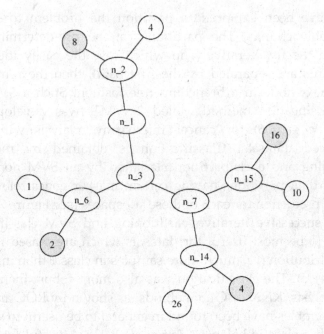

Figure 7-2. Example of an SVM decision tree for a two-class central nervous system cancer dataset classification by microarray data. Ellipses are successively numbered decision nodes The two classes are in the end nodes, class 1: filled circles, class 2: empty circles.. Numbers inside circles indicate the number of patients classified to the end node. From Pang *et al.* 2006. Reproduced by permission from IEEE, © 2006 IEEE.

information about specific data items that are easy or difficult to classify, which is always useful for discovering new hidden factors in complex biological data.

7.3 Methods for microarray data

A general problem in selecting good small sets of biomarkers from a large set is to achieve a consistently well-performing set of markers which is of minimum size. This issue is independent of the classification method itself used to evaluate the worth of individual markers or their interactions, which could be achieved by SVM, graph based methods and many others. Recent approaches to consistency of biomarker sets are discussed in subchapter 7.3.1.

As for the classification techniques themselves, Support Vector Machines are now routinely used to analyse biomarker data with good success in an unsupervised manner, and will not be discussed in detail here. An example of SVM trees used for cancer microarray data was already presented above in section 7.2. Other types of classification methods, however, deserve attention, especially semisupervised methods.

Semisupervised methods are based on the notion that unlabeled data are also used, in addition to the known labeled training data. A good introduction and review of semisupervised methods is in Zhu (2008). Such methods are being taken up in the bioinformatics field also. It is interesting to note the human brain appears to perform classification in a manner similar to EM algorithm for a Gaussian mixture model (Zhu *et al.* 2007). The benefit of using unlabeled data for model parameter estimation is evident in the following Figure 7-3.

According to Zhu (2008), if one is already using SVM for microarray data, transductive SVM (TSVM) is a good way to try to use more of available data, because the additional data points help to put the decision boundary into an area sparse with data points, leading to less misclassifications. On the other hand, if samples with similar features seem to belong to same classes, graph-based semisupervised methods may be best. This may hold especially for microarray type data, where expression signature similarity is taken to mean similarity in gene function and involvement in certain metabolism reactions. This is equivalent to the cluster assumption as explained in the book of Chapelle *et al.* (2006).

Graph based methods appear also attractive, because they can incorporate hundreds of thousands of features in the analysis, as indicated in Hwang *et al.* (2008). They developed a semi-supervised method based on network propagation by a spectral graph transducer, in which bipartite graphs are used to analyse data. They introduced a more efficient network propagation

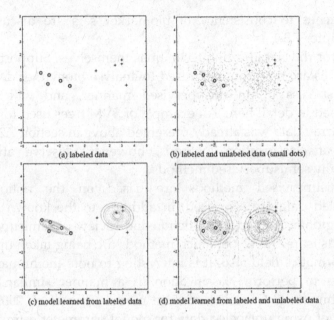

(a) labeled data (b) labeled and unlabeled data (small dots)

(c) model learned from labeled data (d) model learned from labeled and unlabeled data

Figure 7-3. In semisupervised learning, unlabeled data can improve classification modelling markedly (Zhu 2008). Reproduced by permission of the author.

algorithm to increase the efficiency of the computation and achieved converging solutions. The new network propagation method performed consistently better than SVM, LDA or Naïve Bayes on microarray gene data. Furthermore, the method seems to be more robust with different datasets.

In general it seems that semisupervised methods should be evaluated further for complex datasets like microarray data to improve classification performance.

7.3.1 *Gene selection algorithms*

Feature selection is an essential part for analysis of high dimensional bioinformatics data in which a large number of variables is available as explaining variables for clustering or prediction. A good review on the different kinds of feature selection methods from data mining perspective is in Saeys *et al.*

(2007), with a good list of algorithms for microarray data analysis. It appears that robustness of feature selection should also be considered in selecting the best classification model, in addition to mere model performance. Different datasets may actually need different types of feature selection for the most consistent classification (Saeys *et al.* 2007, 2008).

The robustness of a classification method can be measured by the average over all pairwise similarity comparisons between the different feature selection methods. Saeys *et al.* found that in general ensemble feature selection methods are more robust (*i.e.* performance is more stable when changes are made to the analysed dataset). Other possible ways of assessing (and choosing the best) robustness of classification are discussed elsewhere in this book (see subchapter 7.3.2).

The repeatability (consistency) of the algorithm to yield the same result (*i.e.* deterministic nature of the algorithm) is also desirable, especially if the results are to be used for medical diagnosis. Basically it is a matter of good feature selection techniques, which are usefully reviewed by Saeys *et al.* (2007).

Most methods for gene selection are based on estimating class-separability directly, but do not consider reproducibility on new data. Such methods select genes iteratively by bootstrapping. These methods include the hybrid KNN/GA method of Li *et al.* (2001) and the LDA/GA wrapper method of Huerta *et al.* (2008, reviewed below). A general review of these methods is in Wahde & Szallasi (2006). Though often finding good gene sets, these methods suffer from poor repeatability and are difficult to set up with suitable fitness functions and stopping criteria for the Genetic Algorithm part.

Huerta *et al.* (2008) developed a hybrid wrapper method using LDA to select a set of informative genes in a first step, then using a Genetic Algorithm to modify the gene set, not by random recombination, but by a kind of "directed evolution" informed by the LDA parameters (LDA-based cross-over of the "chromosomes" in GA), and repeating the process until a good solution is found. Mutations remove poorly-performing genes in each generation. The classification accuracies were better than in 16 other recent algorithms, including the consistency bootstrapping approach of Pang *et al.* (2007, discussed below). However, because Huerta *et al.* (2008) used 10-fold crossvalidation, results are not clearly comparable with the consistency approach of Pang *et al.* (2007).

It would be interesting to see if the good results could be replicated consistently by running the LDA-GA analysis say 10 times to find out, whether the classification accuracies would have been the same and the number and identity of selected genes the same (*i.e.* a deterministic solution reached consistently). If this would be so, then the method of Huerta could truly be said to have found the best consistent solution reported so far in the literature in the studied 7 cancer datasets. What is surprising in their results is the small number of genes which achieve such good performance. Replication of similar results on other datasets is awaited with interest. The good results may be explained in part by the fact that the algorithm explored a larger number of gene candidates (starting with 100 top ranked genes) than the other mainly filter-based methods, which are often limited to the initial top 30 or so genes in the evaluation process.

A new type of integrated method for gene selection was recently introduced by Duval *et al.* (2009), which used integrated SVM selection of gene subsets and ranking of genes with Genetic Algorithm of generating gene candidates. SVM improves the Genetic Algorithm by local search for each selected individual before the production of the next generation population.

7.3.2 *Gene selection by consistency methods*

As indicated in the introduction to this chapter, consistency of performance by the selected set of genes from microarray data for diagnosis of disease using different datasets is of crucial importance for practical applications to determine the disease status of new patients. Many previous publications have been criticized in this respect, including problems in proper validation procedures (Ranhosoff 2005a, 2005b).

Until very recently, the consistency aspect in microarray gene selection from a methodological point of view has attracted little attention. Mukherjee & Roberts (2004, 2005) defined reproducibility by probabilistic consistency as the number of common genes obtained over a pair of subsets randomly drawn from a microarray dataset under the same distribution. A fixed number of genes are selected repeatedly over various subsets of data and the overlap (number and ranking) of common genes is evaluated as a measure of reproducibility (consistency). This approach, however, does not directly address the consistency of results from different subsets of data and may be affected by the order of subsets of data being sampled.

The training set is always a more or less biased subset of the whole population of gene expression samples from all kinds of individuals, in disease prognosis, for example. To address this, Pang *et al.* (2007) proposed a performance-based consistency method with bootstrapping, which can achieve both good class-separability and good reproducibility. In the bootstrapped method the data is partitioned to random subsamples which serve as independent non-overlapping test and training samples. Then one gene is picked at random from each of the training samples. Consistency is calculated by comparing the performance result on training and testing sets, including the case where the training and testing sets are switched. The gene selection process itself can be any method, but was selected to be a genetic algorithm in the papers of Pang *et al.* (2007, 2008). The method was used on a standard set of cancer datasets and obtained superior results compared to previously published methods. For some datasets, poor reproducibility was noted, casting doubt to some earlier published good classification results. These results are considered to be entirely unbiased, because the training and testing sets are kept completely separate until the last performance comparison step (see Figure 7-4).

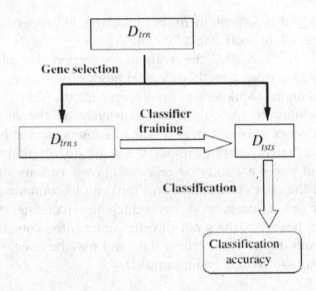

Figure 7-4. Training dataset is divided into separate partitions for training and testing for determining unbiased classification accuracy. Adapted from Pang *et al.* (2007). Reproduced by permission from Springer Science + Business Media.

Consistency (reproducibility) analysis is thus seen as an essential part of gene selection analysis. In the same vein, Elo *et al.* (2008, 2009) recently developed Reproducibility-Optimized Test Statistic (ROTS) for microarray gene selection, in which the gene overlap of sampling is also evaluated, but the number of genes selected is optimized for each data set by measuring reproducibility by the average overlap of top-ranked genes over pairs of boostrapped datasets. Further development of the consistency aspect of gene selection is expected in the near future.

7.4 Genome as a time series and discrete wavelet transform

Genomes and any suitably long strings of DNA can be treated as time series. In this way the traditional time series analysis methods from finance and digital signal processing can be used, after suitable transformation of the DNA sequence to a quantitative two-dimensional time series. Many different transformations can be used, but a simple popular one is as follows (Shieh & Keogh 2008):

If nucleotide = A, then Ti+1 = Ti + 2
If nucleotide = G, then Ti+1 = Ti + 1
If nucleotide = C, then Ti+1 = Ti – 1
If nucleotide = T, then Ti+1 = Ti – 2

This coding scheme usefully guarantees that the reverse direction complementary strings (including palindromes) produces an identical curve in mirror image, that is, with 100% negative correlation. Thus for example many restriction enzyme sites could be identified on both strands of the DNA by corresponding sections of the time series. A simple way to analyse both strands for *e.g.* reversed direction copies of genes would be to concatenate the positive and negative strand timeseries to one longer series which is double the length of the genome.

Another possibility is to use DNA walks in higher dimensionality space and several ways of doing this are possible. Berger *et al.* (2002) proposed using two dimensions by a combination of real and imaginary positive and negative numbers for the four nucleotides. A general review of the various encoding methods and the various data series analysis methods using Digital Signal Processing techniques is given in Lorenzo-Ginori *et al.* (2009). Only some recent representative and interesting exemplary approaches are discussed here.

The time series can be analysed by any available time series methods, including Fourier analysis, autoregression methods and even methods financial stock market analysis. For example, Time dependent ARMA (autoregression moving average) modelling was used to conclude that the randomness of coding segments of genomes is higher than that of non-coding segments (Zielinski *et al.* 2008). In principle any quantitative physical parameters along the DNA sequence can be used, too, for such analysis. Some of known quantitative physical parameters of DNA strings have already been discussed in chapter 4.3.1.

Varadwaj *et al.* (2009) used DSP methods to extract time series features from Electron-Ion Interaction Potential (EIIP) values along long sequences as input to SVM and successfully classified splice sites in the gene coding segments. Autoregression was used by Zhou *et al.* (2009) to estimate spectrum parameters to detect tandem repeats. In the first step, the spectrogram peaks are used to choose areas for more detailed tandem repeat analysis, thus speeding the detection of repeats for large genomes.

Recently developed new algorithms in time series analysis appear to scale to very large datasets and therefore show promise for large-scale genome analysis as well. For example, iSAX method can index and mine terabyte size time series efficiently (Shieh & Keogh 2008). An 11 million nucleotide sequence (human chomosome 2) could be searched fast by chimpanzee sequences using iSAX (see Figure 7-5).

Especially outlier detection in genomes by the time series analysis appears worth exploring further. Yankov *et al.* (2007) developed a very fast method to find outliers (unusual or discordant sections) in the time series, which requires only two scans of the time series. The unusual segments are defined as

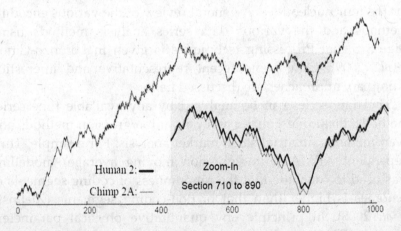

Figure 7-5. Comparison of a segment of human and chimpanzee chromosome 2 using genome coded as time series and iSAX algorithm. From Shieh & Keogh (2008). Reproduced by permission from Springer.

those data items which have the largest distances to their nearest neighbour in the data space. This intriguing method could be used to find unusual genome segments, including unsual repeats or rearrangements or interspecies DNA insertions in arbitrarily large genomes.

Various aspects of Fourier spectral analysis have been used to analyze DNA data, especially Discrete Wavelet Transform (DFT) has been popular to extract good classifier features. For example, Fuentes *et al.* (2008) used a linear combination of three spectral parameters to predict coding regions in genomes. Most methods combine the spectral features with Support Vector Machine processing for classification. Recent successful examples include use of discrete wavelet transform coefficients as input to SVM to analyze MALDI-TOF serum protein profiles of colorectal cancer patients and control healthy patients (Alexandrov *et al.* 2009) and provided both high sensitivity and specificity. Similarly impressive results were obtained by Qiu *et al.* (2009a, 2009b), using DFT data derived from hydrophobicity scales of Chou's pseudo-amino acid composition from G-protein-coupled receptor candidate amino acid sequences as well as protein secondary structure classes. A combination of DFT and SVM could also predict subcellular location of apoptosis proteins (Qiu *et al.* 2009c) and distinguish enzymes from non-enzymes (Qiu *et al.* 2009d).

It seems that if the derived features are quantitative continuous biochemical parameters, the DFT approach can yield good classifying features and thus the time series approach should be more widely considered for proteins with sufficient biochemical characterization.

Another good use case for Fourier and DSP analysis methods appears to be the use of sufficiently large datasets for repeat analysis of genomes. The Fourier analysis can also combined with traditional autocorrelation analysis. Epps (2009) used a hybrid autocorrelation combined with integer period DFT to analyse repeats of tetramer oligonucleotides in the nematode *Caenorhabditis elegans*. The hybrid approach is less likely to yield false positive multiple periodicities often seen in autocorrelation analyses and DFT when used alone.

7.5 Parameterless clustering for gene expression

Gene expression data from microarrays can encompass very large amounts of information from thousands of genes, many tissue or patient samples and long time series datasets from successive measurements. The high dimensionality problems of this kind can be analyzed by parametric methods, which however, demand setting and optimizing various parameters, which can be arbitrary or computationally costly. Therefore various dimensionality reduction methods can be used prior to analysis to reduce the amount of data to be processed, reducing both memory requirement and CPU cycles.

A general review of the many non-linear methods is in van der Maaten *et al.* (2009), which concluded that they often cannot outperform the traditional linear methods, especially the PCA and its improved versions like Rank-based Modified Partial Least Squares (RMPLS) which was found effective for microarray data (Nguyen *et al.* 2009).

Thus the non-linear methods do not appear to have matured yet to be considered generally applicable universal methods. Some algorithms show promise, however, and should be tried to microarray data: t-SNE of van der Maaten & Hinton (2008) and Distance Preserving Dimensionality Reduction DPDR method of Choo *et al.* (2007), as well as autoencoders using neural network learning (Larochelle *et al.* 2009). A great attractive point of DPDR method is that there is no parameter to be optimized to start with.

It would be interesting if these methods were compared side by side for a representative set of microarray datasets. For at least some microarray datasets, IsoMap performs well and can discover biologically relevant groupings (Nilsson *et al.* 2008). Such comparative testing should also include the new kernel PCAs which are showing improving performance.

In microarray analysis a common task is also finding the set of enriched genes in one microarray compared to another one. This has been commonly done using Gene Set Enrichment Analysis (GSEA) of Subramanian *et al.* (2005), but recently a simple univariate approach using regularized t-statistics for permuted data performs surprisingly well (Ackermann & Strimmer 2009). This serves as a warning that simple analysis methods should always be tried first, as they may work very well indeed for many microarray datasets.

7.6 Transductive confidence machines, conformal predictors and ROC isometrics

Standard classification algorithms assign each data item to a single class. In the transductive confidence machine method the classifier can assign more than one class label to a data item, if there is uncertainty about the class to which it belongs. The method is detailed in the book of Vovk *et al.* (2005), and is based on fitting the tested item to all possible classes and approximating the probability of belonging to each class.

A good example of TCM use in bioinformatics is Shahmuradov *et al.* (2005) who classified TATA-motif containing and TATA-less plant promoters by SVM, followed by TCM pruning of items, for which classification should be abstained and using only items with highest credibility. The approach improved promoter prediction and enabled genome-scale scanning of all promoter candidates from the plant *Arabidopsis thaliana*.

TCM based methods with a plugged in SVM and KNN classifier were recently used to analyse cancer microarray data with good results (Yang *et al.* 2009, Wang *et al.* 2009).

Yang *et al.* (2009) developed an alternative approach, Conformer Predictor with Random Forest (CP-RF), which seems even more promising. Conformal predictors are described in Shafer & Vovk (2007), and are based on the predictor giving a p-value for the confidence level of each class prediction. The performance of the classifier can be set beforehand to a desired level and the method is claimed to be well-calibrated regardless of the plugged-in classifier (see example in Figure 7-6). In comparing TCM-SVM, TCM-KNN and CP-RF, the CP-RF showed clearly best and consistent performance at different confidence levels.

However, a direct comparison of CP method with TCM method using the same classifier is necessary to judge which type of hedging classifier performance is actually superior for some/all kinds of datasets. A specific non-conformity measure needs to be developed for each classifier, so that fair comparison between CP and TCM methods may be difficult, if a non-optimal non-conformity measure is used.

The conformal prediction method for multiclass labelling developed by Yang *et al.* (2009) is also interesting, because it can be implemented with specific costs of making certain wrong

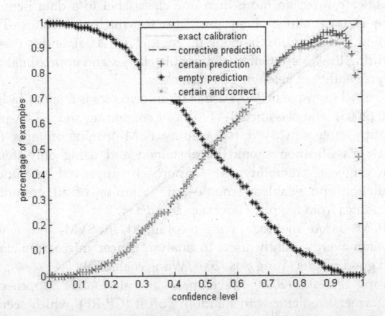

Figure 7-6. Performance of a Conformal predictor with Random Forest (CP-RF) on a subset of chronic gastritis data. Reproduced from Yang *et al.* (2009) by Creative Commons Attribution License from an Open Access article.

predictions, so that the outcomes in practice, *e.g.* medical prediction of dangerous or benign diseases can be weighted appropriately.

The other recent major method of constructing reliable classifiers is ROC isometrics (Vanderlooy *et al.* 2009). ROC (Receiver Operating Characteristic) analysis in general is well-known and established also in bioinformatics data analysis results. The ROC curve displays for the relationship between the probabilities of false positives and false negatives, and has been recently reviewed for biodata analysis use (Lasko *et al.* 2005, Sonego *et al.* 2008). ROC curves and Area Under Curve (AUC) are both already commonly used to compare and select bioinformatics classifiers. For example, Ma & Huang (2005) used maximization of the area under ROC curve to select best set of microarray based gene markers for disease diagnosis. Recently, a handy software tool and web server for statistical comparison of ROC curves was developed to make comparison of classifiers easier (Vergara *et al.*, 2008).

However, ROC method is based on merely two classes, with inevitable misclassifications of both type I and type II errors (false positives and false negatives). ROC isometrics is a way of abstaining from classifying a portion of data items to increase classifier performance for the remaining items to a desired pre-set level. This is achieved by defining isometric lines in ROC space, which determine the optimum classifier with preset proportion of false positive and false negatives (Figure 7-7).

ROC isometrics and TCM were used to improve the performance of predictors for plant polyadenylation sites from genomic sequences by HIKs (Highly informative K-mers). Classification by Linear Discriminant Classifier (LDC), followed by TCM or ROC isometrics, was equally efficient in improving classifier performance (Figure 7-8) by reliably identifying the items, which are difficult to classify.

ROC isometrics guarantees the minimum number of items to be left unclassified to obtain the desired classifier performance. An

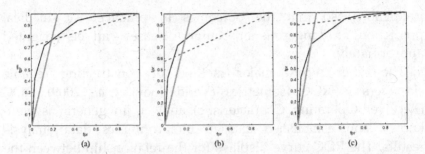

Figure 7-7. ROC isometrics. In case (c), a number of data items have to be left unclassified to obtain the desired performance. X-axis: false positive rate, Y-axis: true positive rate. Reproduced from Havukkala & VanderLooy (2007) by author's right for personal reuse.

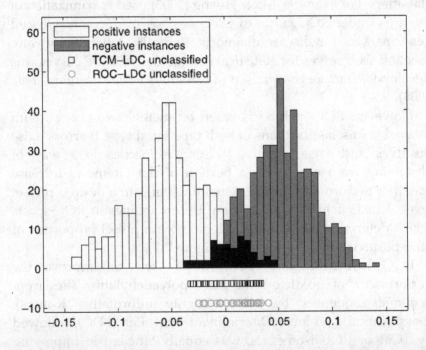

Figure 7-8. Abstaining from classifying certain items obtained by TCM or ROC isometrics. *Arabidopsis* plant genome polyadenylation sites were classified by LDC and the items to be left unclassified to guarantee a specific performance (in this example 99%) correspond well to the overlap between the two classes (black area in the histogram). Reproduced from Havukkala & VanderLooy (2007) by author's right for personal reuse.

additional attractive feature is that the relative importance of false positives and false negatives can be taken into account by weighting them appropriately. In addition, having an objective method to pick out difficult to classify cases for further detailed investigation is very useful, and may even lead to a discovery of more classes to be further classified with another additional classifier.

7.7 Text compression methods for biodata analysis

Many types of compression methods have been used for various kinds of text data storage and indexing for fast retrieval, but only recently have such methods been used for bioinformatics analysis needs. The research area is still fragmented, though a recent general review of the variety of approaches is in Giancarlo *et al.* (2009). This research area shows promise, and includes the Burrows–Wheeler transform and self-indexed transforms discussed further in chapter 9.

Text compression *per se* is useful for reducing storage and transfer costs of large amounts of genomic data, and the first practical systems are maturing. Christley *et al.* (2009) achieved 780-fold compression of the human genome from 3.1 Gigabytes to 4.2 Mb and Brandon *et al.* (2009) obtained 433-fold text compression of a set of mitochondrial genomes. Such methods are very helpful, but still need refinements and standardization for wide-spread adoption. It also may be that genomic data collections with different amounts of mutation rate and repetitiveness need customised solutions for maximum efficiency.

Kolmogorov complexity is a measure of compressibility of any symbol sequence. It can be utilized for comparing two sequences by obtaining relative compressibility of two symbol strings alone or together, with the idea that if two sequences compress better concatenated together than singly, they must be somehow related. This simple idea can be implemented in many ways using various kinds of compressibility indexes and it is not yet clear, which of these are most suitable for *e.g.* DNA and protein sequence analyses. It seems, however, that at least compared to BLAST and Smith-Waterman, no improved accuracy of finding related sequences is achieved, as already discussed in subchapter 4.2.2.

For other bioinformatics applications, the Kolmogorov complexity has not been widely adopted yet and its superiority compared to other methods needs to be more convincingly demonstrated to obtain mainstream method status. Recent examples of using Kolmogorov complexity for biodata analysis include the work of Apostolico *et al.* (2006), who mined for protein motifs by compressing them. Gilbert *et al.* (2007) compared Topology of Protein Structure (TOPS) strings, and further work in this direction is progressing (Liu & Wang 2008). Another interesting application domain is microRNA structure search by data compressibility (Evans *et al.* 2007). It is a welcome addition to researchers hunting for polymorphism in microRNAs, which are best studied with a combination of different algorithms for structural and functional similarity in the important area of pharmacogenomics (Mishra & Bertino 2009).

Text compression can also be achieved by elucidating a context-free grammar which generates a text string sufficiently similar to the original string. The grammar rules then encode in compressed form the original string. RNACompress of Liu *et al.* (2008) represents the RNA structure as grammar rules and also show two orders of magnitude improvement in compression and decompression speed compared to previous methods. Interestingly, the Kolmogorov complexity of the RNA structure correlated with its enzymatic activity, suggesting that evolution in general refines the RNAs by increasing the informativeness of the molecule to perform a complicated task.

In summary, data compression as biodata analysis tool is a heterogeneous field, as no unifying approaches have been identified. However, machine learning experts are already utilizing data compression as a tool for selecting best classifiers, and feature selection in the broad sense and SVM projection of data to higher dimensional space are also forms of data compression/reduction, which are used extensively in bioinformatics also. Further expanded use of text compression methods in genomic data analysis is to be expected, but are likely to be customised solutions to different types of data, due to the varying informativeness, complexity and repetitiveness of different genome and proteome datasets.

References

Ackermann, M. & Strimmer, K. 2009. A general modular framework for gene set enrichment analysis. *BMC Bioinformatics* 10: 47.

Alexandrov, T., Decker, J., Mertens, B., Deelder, A.M., Tollenaar, R.A., Maass, P. & Thiele, H. 2009. Biomarker discovery in MALDI-TOF serum protein profiles using discrete wavelet transformation. *Bioinformatics* 25(5): 643–649.

Apostolico, A. *et al.* (2006) Mining, compressing and classifying with extensible motifs. *Alg. Mol. Biol.*, 1, 4.

Ben-Hur, A., Ong, C.S., Sonnenburg, S., Schölkopf, B. & Rätsch, G. 2008. Support Vector Machines and Kernels for Computational Biology. *PLoS Comput Biol* 4(10): e1000173.

Berger, J.A., Mitra, S.K., Carli, M. & Neri, A. 2002. New approaches to genome sequence analysis based on digital signal processing. *Paper presented in the Proceedings of the Workshop on Genomic Signal Processing and Statistics, Raleigh, NC, USA.*

Borgwardt, K.M., Ong, C.S., Schönauer, S., Vishwanathan, S.V.N., Smola, A.J. & Kriegel, H.-P. 2007. Protein function prediction via graph kernels. *ISMB (Supplement of Bioinformatics) pp. 47–56.*

Brandon, M.C., Wallace, D.C. & Baldi, P. 2009. Data structures and compression algorithms for genomic sequence data. *Bioinformatics* 25(14): 1731–1738.

Bunke, H. & Riesen, K. 2008. A Family of Novel Graph Kernels for Structural Pattern Recognition. *Lecture Notes in Computer Science* 4756: 20–31.

Chapelle, O., Schölkopf, B. & Zien, A. (eds) 2006. Semi-Supervised Learning. *MIT Press, Cambridge, MA.*

Choo, J., Kim, H., Park, H. & Zha, H. 2007. A Comparison of Unsupervised Dimension Reduction Algorithms for Classification. *IEEE International Conference on Bioinformatics and Biomedicine BIBM 2007, pp. 71–77.*

Chow, R., Zhong, W., Blackmon, M., Stolz, R. & Dowell, M. 2008. An efficient SVM-GA feature selection model for large healthcare databases. *Proceedings of the 10th annual conference on Genetic and evolutionary computation, pp. 1373–1380.*

Christley, S., Lu, Y., Li, C., Xie, X. 2009. Human genomes as email attachments. *Bioinformatics.* 25(2): 274–275.

Duval, B., Hao, J.K. & Hernandez, J.C. 2009. A Memetic Algorithm for Gene Selection and Molecular Classification of Cancer. *ACM conference proceedings, GECCO'09, July 8–12, 2009, Montréal Québec, Canada.*

Elo, L.L., Filen, S., Lahesmaa, R. & Aittokallio, T. 2008. Reproducibility-optimized test statistic for ranking genes in microarray studies. *IEEE/ACM Transactions on Computational Biology and Bioinformatics 5(3): 423–431.*

Elo, L.L., Hiissa, J., Tuimala, J., Kallio, A., Korpelainen, E., Aittokallio, T. 2009. Optimized detection of differential expression in global profiling experiments: case studies in clinical transcriptomic and quantitative proteomic datasets. *Brief Bioinform. 2009 Sep;10(5):547–55. Epub 2009 Jun 23.*

Epps, J. 2009. A Hybrid Technique for the Periodicity Characterization of Genomic Sequence Data. *EURASIP Journal on Bioinformatics and Systems Biology 2009: Article ID 924601, pp. 1–8.*

Evans, S.C. *et al.* 2007. MicroRNA target detection and analysis for genes related to breast cancer using MDLcompress. *EURASIP J. Bioinform. Syst. Biol., 2007, 1–16.*

Fehr, J., Zapién Arreola, K. & Burkhardt, H. 2008. Fast Support Vector Machine Classification of Very Large Datasets. *Pp. 11–18 in: Proceedings of the 31st Annual Conference of the Gesellschaft für Klassifikation e.V., Albert-Ludwigs-Universität Freiburg, March 7–9, 2007.*

Fuentes, A.R., Ginori, J.V.L. & Ábalo, R.G. 2008. A New Predictor of Coding Regions in Genomic Sequences using a Combination of Different Approaches. *International Journal of Biological and Medical Sciences 3(2): 106–110.*

Gao, Y., Sprecher, A.J. & Jiang, J.J. 2009. Support Vector Machine Based Decision Tree to Classify Voice Pathologies Using High-Speed Videoendoscopy. *Unpublished manuscript.*

Gehler, P. V. & S. Nowozin: Let the Kernel Figure it Out: Principled Learning of Pre-processing for Kernel Classifiers. *Proceedings of the IEEE Computer Society Conference on Computer Vision and Pattern Recognition (CVPR 2009), 1–8 (accepted) (06 2009).*

Giancarlo, R., Scaturro, D. & Utro, F. 2009. Textual data compression in computational biology: a synopsis. *Bioinformatics 25(13): 1575–1586.*

Gilbert,D. *et al.* (2007) Alignment-free comparison of TOPS strings. *Proceedings of London Algorithmics and Stringology. College Publications, pp. 177–197.*

Gönen, M. & Alpaydin, E. 2008. Localized multiple kernel learning. *In Proceedings of the 25th international Conference on Machine Learning (Helsinki, Finland, July 05 - 09, 2008). ICML '08, 307: 352–359.*

Havukkala, I. & Vanderlooy, S. 2007. On the Reliable Identification of Plant Sequences Containing a Polyadenylation Site. *Journal of Computational Biology. 14(9): 1229–1245.*

Huerta, E.B., Duval, B. & Hao, J.K. 2008. Gene Selection for Microarray Data by a LDA-Based Genetic Algorithm. *Proceedings of the Third IAPR International Conference, Springer.*

Hwang, T., Sicotte, H., Tian, Z., Wu, B., Kocher, J.P., Wigle, D.A., Kumar, V. & Kuang, R. 2008. Robust and efficient identification of biomarkers by classifying features on graphs. *Bioinformatics 24(18): 2023–2029.*

Jacob, L. & Vert, J-P. 2007. Kernel methods for in silico chemogenomics. *Technical Report 0709.3931v1,arXiv, 2007.*

Jacob L. & Vert, J.-P. 2008. Protein-ligand interaction prediction: an improved chemogenomics approach. *Bioinformatics, 24(19): 2149–2156.*

Kashima H., Idé, T, Kato T. & Sugiyama M. 2009. Recent Advances and Trends in Large-scale Kernel Methods. *IEICE Transactions on on Information and Systems, vol.E92-D, in press.*

Kloft, M, Brefeld, U., Laskov, P. & Sonnenburg, S. 2008. Non-sparse Multiple Kernel Learning. *In: Twenty-Second Annual Conference on Neural Information Processing Systems NIPS08, Workshop: Kernel Learning: Automatic Selection of Optimal Kernels.*

Larochelle, H., Bengio, Y., Louradour, J. & Lamblin, P. 2009. Exploring strategies for training deep neural networks. *Journal of Machine Learning Research, 10: 1–40.*

Lasko, T.A., Bhagwat, J.G., Zou, K.H. & Ohno-Machado, L. 2005. The use of receiver operating characteristic curves in biomedical informatics. *J. of Biomedical Informatics 38: 404–415.*

Li, L., Weinberg, C.R. *et al.* 2001. Gene selection for sample classification based on gene expression data: study of sensitivity to choice of parameters of the GA/KNN method. *Bioinformatics 17(12): 1131–1142.*

Lippert, C., Stegle, O., Ghahramani, Z. & Borgwardt, K. 2009. A kernel method for unsupervised structured network inference. *Pp. 368–375 In: Proceedings of the 12th International Conference on Articial Intelligence and Statistics (AISTATS) 2009, Clearwater Beach, Florida, USA. Volume 5 of JMLR: W&CP 5.*

Liu, L. & Wang, T. 2008. Comparison of TOPS strings based on LZ complexity. *Journal of Theoretical Biology 251(1): 159–166.*

Liu, Q. *et al.* 2008. RNACompress: grammar-based compression and informational complexity measurement of RNA secondary structure. *BMC Bioinformatics, 9: 176.*

Lorenzo-Ginori, J.V., Rodriguez-Fuentes, A., Abalo, R.G. & Rodriguez, R.S. 2009. Digital Signal Processing in the Analysis of Genomic Sequences. *Current Bioinformatics 4(1): 28–40.*

Ma, S. & Huang, J. 2005. Regularized ROC method for disease classification and biomarker selection with microarray data. *Bioinformatics 21(24): 4356–4362.*

Mishra, P.J. & Bertino, J.R. 2009. MicroRNA polymorphisms: the future of pharmacogenomics, molecular epidemiology and individualized medicine. *Pharmacogenomics 10(3): 399–416.*

Morita, K., Saito, Y., Sato, K., Oka, K., Hotta, K., Sakakibara, Y. 2009. Genome-wide searching with base-pairing kernel functions for noncoding RNAs: computational and expression analysis of snoRNA families in Caenorhabditis elegans. *Nucleic Acids Res 37(3): 999–1009.*

Mukherjee, S. & Roberts, S.J. 2004. Probabilistic Consistency Analysis for Gene Selection. *In: Proceedings of the 2004 IEEE Computational Systems Bioinformatics Conference, pp. 487–488.*

Mukherjee, S., Roberts, S.J., van der Laan, M.J. 2005. Data-adaptive test Statistics for microarray data. *Bioinformatics 21: ii108–14.*

Mundra, P.A. & Rajapakse, J.C. 2007. SVM-RFE with Relevancy and Redundancy Criteria for Gene Selection. *Lecture Notes in Computer Science 4774: 242–252.*

Nguyen, T.S. & Rojo, J. 2009. Dimension Reduction of Microarray Data in the Presence of a Censored Survival Response: A Simulation Study. *Statistical Applications in Genetics and Molecular Biology: 8(1): Article 4.*

Nilsson, J., Fioretos, T., Höglund, M. & Fontes, M. 2008. Approximate geodesic distances reveal biologically relevant structures in microarray data. *Bioinformatics 20(6): 874–880.*

Pang, S., Havukkala, I. & Kasabov, N. 2006. Two-class SVM trees (2-SVMT) for biomarker data analysis. Lecture Notes in Computer Science (LNCS) 3973: 629–634.

Pang, S. & Kasabov, N. 2008. Encoding and decoding the knowledge of association rules over SVM classification trees. *Knowledge and Information Systems 19(1): 79–105.*

Pang, S., Havukkala, I., Hu, Y. & Kasabov, N. 2007. Classification consistency analysis for bootstrapping gene selection. *Neural Computing and Applications 16(6): 527–539.*

Pang, S., Havukkala, I., Hu, Y. and Kasabov, N. 2008. Bootstrapping Consistency Method for Optimal Gene Selection from Microarray Gene Expression Data for Classification Problems. *Chapter 11, pp. 89–110 in: Y.-Q. Zhang and Jagath C Rajapakse (eds.), "Machine Learning in Bioinformatics," Wiley Book Series on Bioinformatics: Computational Techniques and Engineering, John Wiley & Sons.*

Qiu, J-D., Huang, J-H., Liang, R-P. & Lu, X-Q. 2009a. Prediction of G-protein-coupled receptor classes based on the concept of Chou's pseudo amino acid composition: An approach from discrete wavelet transform. *Analytical biochemistry 390(1): 68–73.*

Qiu, J-D., Luo, S-H., Huang, J-H., Liang, R-P. 2009b. Using support vector machines for prediction of protein structural classes based on discrete wavelet transform. *Journal of Computational Chemistry 30(8), 1344–1350.*

Qiu, J-D., Luo, S-H., Huang, J-H., Sun, X-Y. & Liang, R-P. 2009c. Predicting subcellular location of apoptosis proteins based on wavelet transform and support vector machine. *Amino Acids (in press).*

Qiu, J-D., Luo, S-H., Huang, J-H. & Liang, R-P. 2009d. Using support vector machines to distinguish enzymes: Approached by incorporating wavelet transform. *Journal of Theoretical Biology 256(4): 625–631.*

Ranhosoff, D.F. 2005a. Bias as a threat to the validity of cancer molecular-marker research. *Nat Rev Cancer 5(2): 142–149.*

Ranhosoff, D.F. 2005b. Lessons from controversy: Ovarian cancer screening and serum proteomics. *Journal of National Cancer Institute 97(4): 315–319.*

Saeys, Y., Abeel, T., Van de Peer, Y. 2008. Towards robust feature selection techniques. *Proceedings of Benelearn 2008, pp. 45–46.*

Saeys, Y., Inza, I. & Larrañaga, P. 2007. A review of feature selection techniques in bioinformatics. *Bioinformatics 23(19): 2507–2517.*

Sakakibara, Y., Popendorf, K., Ogawa, N., Asai, K., Sato, K. 2007. Stem kernels for RNA sequence analyses. *J Bioinform Comput Biol 5(5):1103–1122.*

Sato, K., Mituyama, T, Asai, K. & Sakakibara, Y. 2008. Directed acyclic graph kernels for structural RNA analysis. *BMC Bioinformatics 9: 318.*

Schölkopf, B., Tsuda, K. and Vert, J.P. 2004 Kernel Methods in Computational Biology. *MIT Press, Cambridge, MA.*

Schultheiss, S.J., Busch, W., Lohmann, J.U., Kohlbacher, O., Rätsch, G. 2009. KIRMES: Kernel-based identification of regulatory modules in euchromatic sequences. *Bioinformatics 25(16): 2126–2133.*

Shafer, G. & Vovk, V. 2007. A tutorial on conformal prediction. *J Machine Learning Research 9: 371–421.*

Shahmuradov, I. A., Solovyev, V. V. & Gammerman, A. J. 2005 Plant promoter prediction with confidence estimation. *Nucleic Acids Research 33(3): 1069–1076.*

Shawe-Taylor, J. & Cristianini, N. 2004. Kernel Methods for Pattern Analysis. *Cambridge Univ. Press.*

Shieh, J. & Keogh, E. 2008. iSAX: Indexing and Mining Terabyte Sized Time Series. *SIGKDD 2008.*

Sonego, P., Kocsor, A. & Pongor, S. 2008. ROC analysis: applications to the classification of biological sequences and 3D structures. *Brief Bioinform. 9(3):198–209.*

Sonnenburg, S., Raetsch, G. & Rieck, K. 2007. Large scale learning with string kernels. *In: Large Scale Kernel Machines (2007) MIT Press, Cambridge, MA, pp. 73–103.*

Sonnenburg, S., Zien, A. &. Rätsch, G. 2006. ARTS: Accurate Recognition of Transcription Starts in Human. *Bioinformatics, 22(14): e472–e480.*

de Souto, M.C., Prudencio, R.B.C., Soares, R.G.F., de Araujo, D.S.A., Costa, I.G., Ludermir, T.B. & Schliep, A. 2008a. Ranking and Selecting Clustering Algorithms Using a Meta-Learning Approach. *Pp. 3728–3734. In Proc. of IEEE International Joint Conference on Neural Networks. IEEE Computer Society.*

de Souto, M.C., Costa, I.G., de Araujo, D.S., Ludermir, T.B., Schliep, A. 2008b. Clustering cancer gene expression data: a comparative study. *BMC Bioinformatics 9: 497.*

Souza, B., Carvalho, A. & Soares, C. 2008. Meta–learning for gene expression data classification. *Pp. 441–446 in: HIS '08. Eighth International Conference on Hybrid Intelligent Systems.*

Subramanian, A., Tamayo, P., Mootha, V.K., Mukherjee, S., Ebert, B.L., Gillette, M.A., Paulovich, A., Pomeroy, S.L., Golub, T.R., Lander, E.S. & Mesirov, J.P. 2005. Gene set enrichment analysis: a knowledge-based approach for interpreting genome-wide expression profiles. *Proc Natl Acad Sci USA 102: 15545–15550.*

Sun, L., Ji, S. & Ye, J. 2008. Adaptive diffusion kernel learning from biological networks for protein function prediction. *BMC Bioinformatics* 9: 162.

Vanderlooy, S., Sprinkhuizen-Kuyper, I., Smirnov, E. & van den Herik, J. 2009. The ROC Isometrics Approach to Construct Reliable Classifiers. *Intelligent Data Analysis 13(1): 3–37.*

van der Maaten, L.J.P. & Hinton, G.E. 2008. Visualizing data using t-SNE. *Journal of Machine Learning Research,* 9: 2431–2456.

van der Maaten, L.J.P., Postma, E.O. & van den Herik, H.J. 2009. Dimensionality Reduction: A Comparative Review. *Journal of Machine Learning Research, 10:* 1–41.

Varadwaj, P., Purohit, N. & Arora, B. 2009. Detection of Splice Sites Using Support Vector Machine. *In: Book Series Communications in Computer and Information Science, Volume 40, Contemporary Computing, Part 10, pp. 493–502.*

Vergara, I.A., Norambuena, T., Ferrada, E., Slater, A.W. & Melo, F. 2008. StAR: a simple tool for the statistical comparison of ROC curves. *BMC Bioinformatics* 9: 265.

Vert, J-P. & Jacob, L. 2008. Machine Learning for In Silico Virtual Screening and Chemical Genomics: New Strategies. *Combinatorial Chemistry & High Throughput Screening, Vol. 11(8): 677–685.*

Vert, J.-P., Matsui, T., Satoh, S. & Uchiyama, Y. 2009. High-level feature extraction using SVM with walk-based graph kernel. *Proceedings of the IEEE International Conference on Acoustics, Speech and Signal Processing (ICASSP 2009).*

Vovk, V., Grammerman, A., & Shafer, G. 2005. *Algorithmic Learning in a Random World. Springer, New York.*

Wahde, M. & Szallasi, Z. 2006 Improving the prediction of the clinical outcome of breast cancer using evolutionary algorithms. *Soft Comput 10(4): 338–345.*

Wang, H., Lin, C., Yang, F. & Hu, X. 2009. Hedged predictions for traditional Chinese chronic gastritis diagnosis with confidence machine. *Computers in Biology and Medicine 39(5): 425–432.*

Yan, R. Liu, Y. Jin, R. & Hauptmann, A. 2003. On predicting rare classes with SVM ensembles in scene classification. *IEEE International Conference On Acoustics Speech And Signal Processing 3 (III): 21–24.*

Yang, F., Wang, H.Z., Mi, H., Lin, C.D. & Caim, W.W. 2009. Using random forest for reliable classification and cost-sensitive learning for medical diagnosis. *BMC Bioinformatics 10(Suppl 1): S22.*

Yang, J., Li, Y., Tian, Y., Duan, L. & Gao, W. 2008. A New Multiple Kernel Approach for Visual Concept Learning. *Lecture Notes In Computer Science 5371: 250–262.*

Yankov, D., Keogh, E. & Rebbapragada, U. 2007. Disk Aware Discord Discovery: Finding Unusual Time Series in Terabyte Sized Datasets. *ICDM 2007.*

Zaslavskiy, M., Bach, F. & Vert, J.-P. 2009. Global alignment of protein-protein interaction networks by graph matching methods. *Bioinformatics (to appear).*

Zhou, H., Du, L. & Yan, H. 2009. Detection of Tandem Repeats in DNA Sequences Based on Parametric Spectral Estimation. *IEEE Transactions on Information Technology in Biomedicine 13(5): 747–755.*

Zhu, X. 2008. Semi-supervised learning literature survey. *Tech. Report 1530, Department of Computer Sciences, University of Wisconsin at Madison, Madison, WI, 2008.* http://pages.cs.wisc.edu/~jerryzhu/pub/ssl_survey.pdf. Updated in 2008.

Zhu, X., Rogers, T., Qian, R. & Kalish, C. 2007. Humans perform semisupervised classification too. *Twenty-Second AAAI Conference on Artificial Intelligence (AAAI-07).*

Zielinski, J.S., Bouaynaya, N., Schonfeld, D. & O'Neill, W. 2008. Time-dependent ARMA modeling of genomic sequences. *BMC Bioinformatics 9(Suppl 9): S14.*

Chapter 8

Integration of Multimodal Data: Toward Systems Biology

Systems biology is the generic area of analyzing an organism as a whole in terms of its metabolism, gene expression and gene regulation. Under this rubric comes a variety of data analysis approaches with different combinations of multimodal data sets. In modern systems biology, the increasing flood of data gives partial and overlapping information about connections in metabolic and regulatory networks. The challenging task is to screen out the unreliable pieces of data and to connect and integrate the connections and subgraphs of networks to a coherent whole which can be validated and compared with other datasets of network data.

Algorithms for graph comparisons and network matching are important methods in telecommunications, internet networking, computer chip design and so on, and such techniques can be usefully adapted to bioinformatics problems in the systems biology domain. However, in biology such data analysis tasks are often more challenging than in engineering, because the data is more heterogeneous, is collected using a wide variety of experimental or *in silico* methods with varying accuracies, and often even harmonizing the data labeling and metadata between data sources to be integrated poses problems.

Data representation standards for systems biology are essential but are still under development and include Systems Biology Ontology (http://www.ebi.ac.uk/sbo/, ontologies in general are

245

discussed in next chapter). A new notation system is the Systems Biology Graphical Notation (SBGN) of Le Novère *et al.* (2009), which tries to standardize entity relationships, process diagrams and quantitative representation of process flows using standardized visualization icons and ontology standards. Though such standards have yet to be adopted widely for systems biology analysis, there is already a burgeoning literature of integrated analyses of post-genomic data. Recent advanced methods relevant to systems biology are exemplified in this chapter, with pertinent comparisons and pointers to literature.

8.1 Comparative genome annotation systems

Genome comparisons and phylogenomics is one approach to aid systems biology analysis, when single genome data is not sufficient to decipher metabolic pathways or gene regulatory networks. A common large-scale graph analysis task is the linkage of similar DNA or protein sequences by all versus all pair-wise BLAST comparisons or by other similarity search algorithms. The proteome network COGS (Clusters of Orthologous Groups of Proteins) at NCBI was built using exhaustive pair-wise comparisons of proteomes using bidirectional best hits from all major phylogenetic lineages, including Archaea, Bacteria, and eukaryotic plants, fungi and animals (Tatusov *et al.* 2003). This has been extended to a wider compilation of Archaeal and bacterial proteome comparison dataset, called EggNog, with improved similarity criteria and network building methods (Jensen *et al.* 2008).

Graph based methods are based on computing networks of similarity connections between the different sequences and inferring the best clustering of sub-groupings in the network for joining together evolutionarily related sequences. These methods include algorithms like nearest neighbour methods best hit (BeT), reciprocal best hit (RBH), bidirectional best hit (BBH), symmetrical best hit (SymBeT) and reciprocal smallest distance (RSD). The clusters of orthologous group (COGs) is based on BeTs extended by forming congruent triangles of best BLAST hits from at least five different species. These triangles are then merged to larger groups of protein families using single connections. Eukaryotic Gene Orthologs (EGOs) are made in a similar manner for transcribed DNAs. Other graph-based gene and protein family classifications based on graph matching include InParanoid, MultiParanoid, Ortholuge, OrthoMCL, RSD, RoundUp and BUS (referenced in Kuzniar *et al.* 2008).

Hybrid methods include Ensembl Compara, HomoloGene, OrthoParaMap, PhIGs, PHOGs, PhyOP and TreeFam (for references, see Kuzniar *et al.* 2008). These methods are complex systems which merge many kinds of information suitably weighted to get a consensus picture of the evolutionary relationships. They work well in the majority of cases, but may not be best suited to mosaic proteins in which two unrelated proteins have merged during evolution.

Brute-force graph analysis by calculating all versus all similarities can also be used for proteome analysis. Loewenstein *et al.* (2008) successfully produced 1.5 billion BLAST pairwise similarities in a hundred days on the HUGI MOSIX computing grid to produce a comprehensive protein sequence tree for all known proteins. They also discovered many unknown remote homologies between protein families. This represents high-throughput in silico biology delivering unexpectedly useful results. These results will be very useful for reaching the goal of integrative protein function prediction using multiple data sources and algorithms [see the review of Rentzsch & Orengo (2009)].

Automated genome annotation and cross-genome survey of related genes and proteins can also assist system biologists to find functions and regulatory connections to unknown genes. Automatic genome annotation involves multifaceted detection and prediction of a large number of complicated features pertaining to introns, exons, promoters, protein coding regions, UTRs, transcription start sites, mRNA polyadenylation sites, DNA modification sites (*e.g.* methylation sites), repetitive regions and so on. However, gene finding and correct gene structure finding in new genomic sequence data can be challenging.

Conventional gene finding methods are based on Hidden Markov Models trained on carefully validated training data from the target genome. Such HMM methods include GeneMark (Lukashin & Borodovsky 1998), Fgenesh (Salamov & Solovyev 2004), Glimmer-HMM (Majoros *et al.* 2004) and JIGSAW (Allen *et al.* 2006). However, these methods are often based on training data from a single genome and can produce inaccurate prediction on exact borders of predicted regions and even complete false positives and negatives, especially for atypical genes.

Comparative genomics can help, because additional annotations or full length mRNA sequence data from a related genome can be used to improve predictions, as is done by the TwinScan family of software (*e.g.* N-SCAN, Gross & Brent 2006) and many others. The most recent automatic consensus gene structure predictor is EvidenceModeler EVM (Haas *et al.* 2008), which weights the various evidence by type and abundance (e.g. the number of ESTs corresponding to a specific coding region). Benchmarking shows that the accuracy of the system is in most cases superior to previous systems and almost equivalent to human manual annotation. However, the overall accuracy for prediction of genes is still quite low, only about 35%, while the location of exons in found genes is predicted by an accuracy of about 70–80%. The EvidenceModeler is also available on US TeraGrid (www.teragrid.org), in the GenomeGrid subsection for large-scale comparison and analysis of multiple genomes (Gilbert 2008), and has already been used for analysis of multiple Drosophila fly genomes, for example.

The other main type of genomic scale analysis useful for systems biologists is gene expression analysis by comparison of expressed sequence tags (ESTs) to known genes and mRNAs. Large-scale EST sequencing has been routine for a long time and still continues to expand in many areas of biology. The obtained EST data can be used for transcription module and gene expression prediction. For example, PASA software (Haas *et al.* 2003, Haas 2008) integrates existing transcript annotation with EST data to model transcribed sections of genes producing the splice variants best supported by various evidence, including polyadenylation site annotations.

PASA is a complex integrated system incorporating many algorithms, indicating the trend toward large pipeline processing systems having modular architecture and many successive processing algorithms. Such systems are already very powerful for automated analysis of full genomes, provided sufficient additional data (ESTs, protein sequences, phylogenetic comparisons) are available. This enables large-scale *de novo* analysis of novel genomes that keep coming from the increasingly high-throughput sequencing and genome assembly pipelines. Such systems enable even small laboratories to sequence and analyse their favourite genomes and gene expression atlases. A large variety of such pipeline systems already exists, and a good review for such "omics" resources can be found in Vizcaino *et al.* (2009).

8.2 Phylogenetics methods

Deciphering the past ancestral history is also important in understanding evolution and effects of single genes in metabolism for systems biology. Ancestral Recombination Graphs can be calculated in polynomial time from phylogenetic data (Gusfield *et al.* 2004, Semple 2008) and many other methods have been developed for this purpose.

Modern phylogenetic analysis is also already powerful enough to easily detect horizontal gene transfer between unrelated organisms, by using nucleotide composition, oligonucleotide frequencies and codon usage, among others. They are significant for systems biology to pinpoint newly introduced genes and pathways in organisms as well as throwing light into macro-evolutionary events in the distant past. For example, bacterial and

Archaeal common origins were studied by Koonin & Wolf (2008), who posited that there is a very small set of genes (about 70) which are shared among most prokaryotes. Viral evolution is much less known, and viruses, including marine bacteriophages, may actually contain the largest unexplored source of unknown proteins in nature (Edwards & Rohwer 2005). Systems biology analysis of the marine microbiota and virus interactions will be an interesting area in the future, as it affects ocean biology and productivity in a global scale.

Ortholog network algorithms are reviewed by Kuzniar *et al.* (2008). Such methods are based either on tree clustering, graph clustering or hybrid methods. In tree clustering, the sequences are placed a phylogenetic tree by pairwise comparisons of sequences. The tree is constructed either based on a known species tree onto which the new sequences are placed, or the tree is computed *de novo* from the input sequences only, using some heuristic method of evaluating the optimality of the tree, *e.g.* by computing many alternative trees and choosing the most parsimonious solution. These methods rely on the accuracy of multiple alignments as well as the proper criteria of optimality/parsimoniousness, based on a specific evolutionary model of changes in sequences during descent from the common ancestor. Tree clustering methods include COCO-CL, HOPS, LOFT, RAP, SDI and RIO. Unfortunately, tree-based methods do not scale well to large data sets.

A technical note on phylogenetic data formatting before analysis: the currently favoured method in phylogenomics is to concatenate all proteins in each genome before multiple alignment and phylogenetic alignment, instead of using pairwise single proteins aligned to best matchine proteins in the other genomes (orthology trees). Comparison of orthology trees and concatenated proteomes in phylogenomics is given in Dutilh *et al.* (2007). Concatenated proteomes appear to give better results than orthology trees, because consistent trees were obtained for fungi, based on extensive comparison of trees derived from multiple alignments of individual proteins (Marcet-Houben & Gabaldón

2009). However, for prokaryotes which have more horizontal gene transfer, this method may not be so good. Perhaps trying to build a hierarchical tree and force all the data to conform to it is not quite appropriate in cases where horizontal transfer has been significant. Algorithms and methods to determine which one network or tree is a better match to the data still need to be developed.

While phylogeny algorithms comprise a voluminous literature worth a separate book altogether, some less known, but interesting aspects of phylogeny methods are presented below. In general the finding of the optimal tree involves either brute-force enumeration of all possible trees that could fit the data and selecting the best candidate tree(s), or searching the space of possible trees by heuristic sampling. Choosing the most parsimonious solution is called the Minimum Evolution Problem, which is known to be NP-hard. A good recent review of this area is that by Catanzaro *et al.* (2008).

Including more information in building phylogenetic representations of individual genes or DNA or protein motifs could improve both the phylogenetic tree and the multiple alignment. For example, Shih *et al.* (2007) developed a visualization method in which global and local motif logos are displayed next to the phylogeny tree, so that relationships in the evolutionary tree and the motif logos from selected subsets of data can be compared easily. The next step would be to pool the information of phylogenetic relationships to aid in building the correct motifs, in a manner similar to the multiple genome comparative alignment by Paten *et al.* (2009) for ancestral graph resolution.

Sometimes comparing different historical trees is necessary for solid evolutionary analysis. For example, matching a phylogeny tree with biogeographical origin tree can help in deciding whether a biogeographical phylogeny tree is better explained by dispersal between locations followed by speciation or by adaptive radiation in the different locations. A good review of addressing these problems is Ganapathy *et al.* (2006). A related problem is matching several biogeographical origin trees (area cladograms) for different groups of organisms to elucidate the likely biogeographical dispersal history and grouping of the sampled locations. The weighting of the evidence of gene/protein similarities and the absence/presence data in different locations is likely to be crucial in each separate case, depending on the nature of the data.

Another interesting new area is cophylogenic analysis, which studies two organisms (or genes/proteins) which coevolve in dependence with each other. This includes for example symbionts, parasites and plant-herbivore coevolution. Although many methods have been used to compare phylogenetics trees of hosts and parasites or coevolving genes, a formal definition of cophylogenics methods is very recent. Huggins *et al.* (2009) has constructed the first formally defined approaches for creating consistent cophylogenies. An example dataset is shown in Figure 8-1 below. Such analysis could also be extended to coevolving

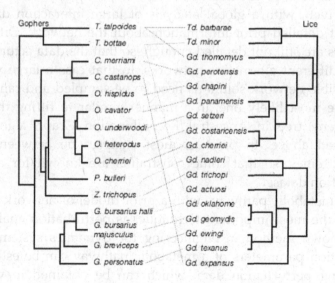

Figure 8-1. A cophylogeny dataset; left, host animal gopher evolution tree; right, louse parasite evolution tree. From Huggins *et al.* (2009). Reproduced from public access preprint archive arXive.

host plants and insects as well, at least in those cases in which the host plants are clearly affected evolutionarily by their herbivore insects, or a specific set of plant genes have evolved to produce anti-herbivore compounds against the herbivores. Provided the methods are sensitive enough, it might even be possible to find out those genes in the plant genome which have been affected by insect herbivores based on the insect-host interaction data fitted to the gene phylogenies of the host plant.

8.3 Network inference from interaction and coexpression data

An important part of systems biology is inference of metabolic pathways and gene regulation networks from large datasets of DNA and protein expression profiles and protein interactions. The analysis can be approached from two directions: top-down and bottom-up. The top-down approach requires some prior model of the network, in which the interactions can then be quantified by perturbation experimental data. The bottom-up approach is more data-driven, with a global analysis of large interaction datasets without detailed prior information about the network. Often the datasets are still not detailed enough, so that the data actually fits many different alternative networks. In that case a large variety of possible network solutions need to be sampled and ranked to find the most likely one, in a manner similar to fitting the best phylogeny tree in Ancestral Recombination Graph algorithms discussed above. In practice, most analyses lie between these two extremes, so that some constraints are known for a large interaction dataset.

For metabolic pathways with a prior model of network nodes, one of the most important techniques is perturbation analysis, a well-known method in engineering for dynamics systems. The interaction parameters of a metabolic pathway can be estimated by using perturbation data, which can be obtained in various ways. One common way to do this is collecting gene expression data from mutants, in which one or more pathways is blocked or changed. However, the estimation of the parameters is difficult with limited data, so that simulation methods are often necessary to fit suitable parameter combinations for an optimal fit to the data. Interestingly, DNA variations among individuals can be regarded as perturbations of a system, to help analyse the network connections (Chen *et al.* 2008), which is a very useful technique, because often making perturbations by experimentally knocking out genes or combinations of two or more genes one by one can be too slow and expensive.

Transcription network analysis is in general based on microarray data on gene and/or protein expression profiles and the correlations between the expression profiles, often added with other types of data on prior knowledge of known interactions and co-occurrence of common regulatory motifs in the promoters of genes coding for co-regulated proteins. One commonly used method is to analyse such gene expression data is Markov Chain Monte Carlo simulation. In short, putative interactions between different nodes in the network are coded in an adjacency matrix as transition probabilities and the iteration of adjacency matrix can be used to model genetic interactions by random walk (Chipman 2009). An example of the use of such a method is the analysis of perturbation data from *Lactococcus lactis* bacterium mutants by Jayawardhana *et al.* (2008).

There is always, however, a danger of over-fitting the data, which can be difficult to assess, if not enough validation data is available. Therefore results of perturbation analysis should be regarded as working hypotheses for experimental validation. Addition of new data may change many details of the inferred network. Here the combination of data from different sources can serve as a powerful validator, especially if the experimental methods are quite different, *e.g.* microarray data on messenger RNA measurements for gene expression and protein binding data for protein interactions.

Another avenue for network analysis is the use of Boolean modelling, Bayesian networks and differential equations or their combinations (discussed further with other methods in next subchapter), but these approaches in general suffer from the problem of having not enough data points to determine the huge number parameters describing the details of the network and the fluxes of regulatory and metabolic proteins. In addition, the quality and reproducibility of microarray and other experimental data is not sufficient for quantitative dynamics simulation of the networks. The huge data space means exploration of the many possible candidate networks to fit the data is still very slow. However, one new method for Bayesian networks based on hierarchical partitioning of the data during analysis (Junga *et al.* 2007)shows a promising speed-up for methods without additional prior knowledge, and such partitioning has already been used in the paper by Sen *et al.* (2009), discussed below.

A way around this computational bottleneck is to avoid determining the whole network dynamics and all connections in detail in one go, but instead trying to look initially for modules, i.e. groups of genes regulated by a common regulator and having similar expression patterns in a specific context or subset of experimental conditions. A gene/protein expression dataset can be divided into subsets or modules having similar expression profiles. This can be achieved by simultaneous clustering of genes and conditions by biclustering methods (Cheng & Church 2000), and such methods have become very popular.

The large variety of biclustering methods already developed is reviewed in Van den Bulcke *et al.* (2006). The best methods allow overlapping modules, so that the algorithm detects the most consistent groupings of expression profiles and not just the best global partitioning of the data. This makes biological sense, because many proteins may participate in several different processes in metabolism and can be differently expressed in different tissues or conditions, cooperating with different sets of genes in different situations. The result is that some genes can appear in more than one module or group of expression profiles. Biologists can then investigate which groupings and functions make sense as working hypotheses and should be further verified by laboratory experiments or further more detailed *in silico* analysis.

After the putative network modules have been defined, the next step is to assemble the whole regulatory network from the modules. The computation of putative gene regulatory connections between the modules (groups of genes) is less difficult than inferring all connections *de novo* from the start, because the genes in one module are assigned a single regulator and the same parameters, which reduces markedly the number of parameters to be estimated from the data.

The assignment of gene modules and the regulatory connections can be achieved simultaneously by the module networks method of Segal *et al.* (2003), in which an iterative Expectation-Maximization (EM) algorithm is used to build a regression tree to determine which regulatory factors best correlate with (= explain) the expression profile of the set of genes in a module (E-step), followed by the reassignment of genes to the module which best fit the current model (M-step). However, Segal's method could be improved by allowing overlapping modules, which is not possible in the original implementation.

Obviously adding more information helps to make the network analysis more reliable. *In silico* analysis can retrieve conserved promoter elements shared by sets of genes, which is suggestive of a common control mechanism, and experimental data can show which transcription factors bind to which gene. Protein–protein interaction datasets are also often available. Such additional data can be included usefully in the integrated approach to network analysis.

In an integrated data analysis approach some of the datasets can be considered as prior knowledge, and used as constraints or starting points of the analysis, *e.g.* as priors in a Bayesian network analysis method (*e.g.* Lee & Lee 2005). Ideally, all of the data should be used simultaneously, and also taking into account the context of the data sets, *e.g.* subsets of data from tissues which have similar regulatory processes. Such truly integrative methods include those of Tanay *et al.* (2002), Wang *et al.* (2002) and Gao *et al.* (2004).

One important and unresolved issue, though, is the weight of evidence to be given to different types of data, which can be difficult. Ideally, the network algorithm itself should assign weights to the different evidence data points, but this is not an easy task without large validation data sets. Thus the true proportions of false positives and false negatives are often unknown.

It is noteworthy, that different methods of gene regulatory network inference can discover different aspects and groupings of related genes. In a recent work using yeast microarray data (Chan *et al.* 2008), Least Angle Regression (LARS) analysis clustered histone regulation related genes, as expected from prior knowledge, and also revealed new gene interactions related to histones that make biological sense, while Kalman Filter analysis with expectation maximization found three groups of cell cycle regulated genes. Evolving Fuzzy Neural Network (EFuNN) found other clusters of interactions and joint analysis by Kalman Filter and EFuNN revealed further candidate regulatory genes. It appears that use and integration of different analysis methods to extract maximum amount of information is beneficial. In addition, it is always good to use various types of data to support each other, *e.g.* gene expression data with protein interaction data and common gene promoter motifs shared by clusters of genes.

However, in spite of the many insights for specific regulatory and interaction subnetwork discoveries, it is somewhat sobering to note that the many sophisticated methods used to analyse well-known large gene expression datasets for gene regulatory network inference show quite different results in the scale of the whole network for the number, size and connectivity of the inferred regulatory modules (see the review of Van den Bulcke *et al.* 2006). There is still a lack of solid validation datasets and standard evaluation criteria to benchmark such methods appropriately. On the positive side, many of the methods when used on real life datasets have already discovered novel regulatory modules which have been verified by independent existing data or by subsequent experimentation and thus contribute significantly to the progress of biology. Many aspects of the genetic regulatory networks which have been reported have to be viewed with caution, but specific regulators and regulatory modules are often reasonable working hypotheses for further investigations and experimental validation by biologists.

Recently Sen *et al.* (2009) developed a context-specific genetic regulatory network inference method in which sets of tissue samples are determined which contain consistent gene expression profiles in the context of the specific set of samples. This partitions the data to subsets that are then analysed separately for gene regulation modules. This approach looks promising, especially because the significance of the contexts (sets of samples) can be statistically evaluated. This method exploits the information of consistent expression profiles in specific subsets of samples to obtain regulatory modules with less noise and could also pinpoint the aberrant tissue samples that do not fit together with other samples from the same tissue. This view of gene modules and interactions in different contexts is in effect a multi-objective analysis (see Handl *et al.* 2007), which is an important new approach in analysing biological networks, because such networks actually serve multiple purposes in nature.

8.4 Bayesian inference, association rule mining and Petri nets

Bayesian inference for gene regulatory networks is a well-established field and has been reviewed recently by Markowetz & Spang (2007). A Bayesian network is a statistical representation of a directed acyclic graph (DAG), with each node described in terms of the node variable and its parent variables in connected nodes in a specific order, which make the node variable independent of the remaining nodes. Defining the network is finding the network which has the highest a posteriori probability given the dataset. A good recent introduction to Bayesian methods in general is in the book of Koski & Noble (2009). In this chapter only recent interesting developments are discussed and compared to alternatives.

Purely Bayesian inference can be improved by incorporating probabilistic models for each data source to be integrated. Such an approach called COGRIM (Chen *et al.* 2007) appears to perform well, seemingly even better than recent association rule based methods like ReMoDiscovery (discussed below). COGRIM integrates three types of data for GRN inference: microarray gene expression data, CHIp-Chip protein interaction data and transcription factor binding site motif data. The probabilistic approach has the advantage that the probability of a regulatory connection can be included in the model, in contrast to considering some network connections to be correct without any weighting by the strength of evidence.

Similar probabilistic versions of Bayesian networks are being developed, *e.g.* Lähdesmäki *et al.* (2008) used principled data fusion by including careful evaluation of reliability of different data sources in the modelling and using Markov Chain Monte Carlo method for exploring the search space. Their method seems to give quite reliable estimates of the probabilities of transcription factor binding to specific promoter sites of different genes compared to traditional promoter scanning methods and COGRIM.

Other hybrid type Bayesian methods keep appearing, but the jury is still out on which of these performs best for which types of data. One notable new semi-supervised Bayesian GRN inference global method called SEREND uses multiple data sources in an iterative classification and logistic regression between targets and regulators (Ernst *et al.* 2008). In another interesting case Bayesian phylogenetic methods were used to analyse even language phylogenies to decipher migrations of human in the Pacific (Gray *et al.* 2009).

An alternative to Bayesian GRN inference is Association Rule Mining (ARM), which is based on exhaustive search of all solutions. The classical ARM method is the Apriori algorithm

(Agrawal *et al.* 1993) which searches for frequent itemsets from bottom up breadth first fashion and performs well in sparse datasets. Another variety is called CHARM (Closed Association Rule Mining) algorithm (Zaki and Hsiao, 2002), which searches the space of possible interactions depth-first.

In the ARM algorithms the data has to be discretized first to a matrix, then frequent itemsets identified from the matrix and then association rules generated. The advantages of the ARM methods include the following: A full search is done, avoiding local maxima, overlapping regulatory modules are possible and efficient pruning of the trees is possible, so that more data can be analysed. The disadvantages are that data discretization may lose or bias the gene expression level information and that often a large number of association rules in generated, requiring careful evaluation and screening for the most significant association rules. In spite of these disadvantages, ARM algorithm called ReMoDiscovery is based on the Apriori approach, is very fast to compute and appears currently competitive with other approaches (Lemmens *et al.* 2006). ARM algorithms are under further development, and seem to be worth following for gene regulatory network analyses as well. An interesting generalized version of association rule mining has been presented recently (Hamrouni *et al.* 2008), and could be used in a systems biology setting, too.

Other notable techniques for biological network analysis include self-organizing maps (SOMs) and Petri nets, used for both gene regulatory networks and metabolic pathways.

As an example, SOMs were used for metabolic pathway analysis by Latino *et al.* (2008) who predicted in genome scale chemical similarities of various metabolic reactions based on MOLMAP metabolic reaction descriptors which encode numerically the chemical structure changes occurring a set of over 4,000 enzymatic reactions. The data was validated with EC enzyme number codes. Another rule-based system SyGMa combined expert knowledge and empirical rules to predict potential metabolites of given parent molecules (Ridder &

Wagener 2008). SOMs have been used occasionally, but their advantages over other network analysis methods have not been properly benchmarked yet for their wider adoption.

Metabolic Petri nets are a network formalism based on token flows from one node (compound) through transitions (biochemical reactions). If the amount of the token (quantity of the chemical) is above a threshold, the transition takes place. Flows of tokens represent the dynamics of chemical flows in the metabolism. Petri nets have been used from late 1990's in systems biology modelling (Hofestädt & Thelen 1998), both for quantitative modelling and for qualitative verification of pathway structure by calculating various invariants of the net (for an example, see Koch *et al.* 2005).

More interesting from a methodological point of view is the inclusion of fuzzy logic into Petri nets, recently suggested by Virtanen (2009). In his method a fuzzy Petri net is a generalization of standard crisp Petri nets. It would be interesting to model biological systems with this method, because in metabolic data there is often uncertainty about the identity of the node (gene, protein), and/or the quantities measured for each node (gene expression, enzyme activity). Another new variation of Petri nets is the Hybrid Functional Petri Nets (HFPN) of Wu & Voit (2009) for modelling metabolic pathways. An in-depth review of this and many other biochemical pathway modelling approaches is available in Chou & Voit (2009).

Finally, at the next complexity level is the problem of comparing two different networks for congruence or crucial differences. The computational cost of matching complex network graphs increases rapidly with network size. Most methods of finding matching networks are heuristic in nature and may not arrive to the best solution. However, one recent method based on matching substructures has been shown to give provably optimal solutions in a reasonable time (Klau 2009), but is still an NP-hard problem. New better ways of comparing and matching complex network structures are still needed to scale up properly to systems biology analyses of postgenomic datasets.

References

Agrawal, R., Imieliński, T. & Swami, A. 1993. Mining association rules between sets of items in large databases. *SIGMOD 22: 207–216.*

Allen, J.E., Majoros, W.H., Pertea, M. & Salzberg, S.L. 2006. JIGSAW, GeneZilla, and GlimmerHMM: puzzling out the features of human genes in the ENCODE regions. *Genome Biol. 7 Suppl 1: S9. 1–13.*

Catanzaro, D., Labe, M., Peseta, R. & Salazar-Gonzalez, J.-J. 2008. Mathematical models to reconstruct phylogenetic trees under the minimum evolution criterion. *Networks 53(2): 126–140.*

Chan, S., Havukkala, I., Jain, V., Hu, Y. & Kasabov, N. 2008. Soft computing method to predict gene regulatory networks: an integrative approach on time-series gene expression data. *Applied and Soft Computing 8(3):1189–1199.*

Chen, G., Jensen, S.T. & Stoeckert, C.J. Jr. 2007. Clustering of genes into regulons using integrated modeling-COGRIM. *Genome Biol. 8(1): R4.*

Chen, Y. *et al.* 2008. Variations in DNA elucidate molecular networks that cause disease. *Nature 452: 429–435.*

Cheng, Y. & Church, G.M. 2000. Biclustering of Expression Data. *Proceedings ISMB, AAAI Press, pp. 93–103.*

Chipman, K.C. & Singh, A.K. 2009. Predicting genetic interactions with random walks on biological networks. *BMC Bioinformatics 10: 17.*

Chou, I.-C. & Voit, E.A. 2009. Recent developments in parameter estimation and structure identification of biochemical and genomic systems. *Mathematical Biosciences 219(2): 57–83.*

Dutilh, B.E., van Noort, V., van der Heijden, R.T., Boekhout, T., Snel, B. *et al.* 2007. Assessment of phylogenomic and orthology approaches for phylogenetic inference. *Bioinformatics 23: 815–824.*

Edwards, R.A. & Rohwer, F. 2005. Viral metagenomics. *Nat. Rev. Microbiol. 3: 504–510.*

Ernst, J., Beg, Q.K., Kay, K.A., Balazsi, G., Oltvai, Z.N. & Bar-Joseph Z. (2008). A semisupervised method for predicting transcription factor-gene interactions in Escherichia coli. *PLoS Comput Biol 4: e1000044.*

Ganapathy, G., Goodson, B., Jansen, R., Le, H., Ramachandran, V. & Warnow, T. 2006. Pattern Identification in Biogeography. *IEEE/ACM Trans. Comput. Biol. Bioinformatics 3(4): 334–346.*

Gao, F., Foat, B.C. & Bussemaker, H.J. 2004. Defining transcriptional networks through integrative modeling of mRNA expression and transcription factor binding data. *BMC Bioinformatics 5: 31.*

Gilbert, D. 2008. A Genome Grid for finding new bug genes. *ISMB poster, 2008.* http://insects.eugenes.org/species/about/genome-grid-ismb08-poster.pdf

Gray, R.D., Drummond, A.J. & Greenhill, S.J. 2009. Language Phylogenies Reveal Expansion Pulses and Pauses in Pacific Settlement. *Science 323(5913): 479–483.*

Gross, S.S. & Brent, M.R. 2006. Using multiple alignments to improve gene prediction. *J Comput Biol.* 13: 379–393.

Gusfield, D., Eddhu, S. & Langley, C. 2007. Optimal, efficient reconstruction of phylogenetic networks with constrained recombination. *J. Bioinformatics and Computational Biology*, 2: 173–213.

Haas, B.J., Delcher, A.L., Mount, S.M., Wortman, J.R., Smith Jr, R.K., Jr., Hannick, L.I., Maiti, R., Ronning, C.M., Rusch, D.B., Town, C.D. *et al.* 2003. Improving the Arabidopsis genome annotation using maximal transcript alignment assemblies. *Nucleic Acids Res*, 31: 5654–5666.

Haas, B.J. 2008. Analysis of Alternative Splicing in Plants with Bioinformatics Tools. *Chapter in: Nuclear pre-mRNA Processing in Plants.*

Hamrouni, T. Ben Yahia, S. & Mephu Nguifo, E. 2008. GARM: Generalized association rule mining. *In: Proceedings of the 6th International Conference on Concept Lattices and their Applications (CLA 2008), pp. 145–156.*

Handl, J., Kell, D.B. & Knowles, J. 2007 Multiobjective Optimization in Bioinformatics and Computational Biology. *IEEE/ACM transactions on computational biology and bioinformatics, 4(2): 279–292.*

Hofestädt, R. & Thelen, S. 1998. Quantitative modelling of biochemical networks. *In: Silico Biol.* 1: 39–53.

Huggins, P., Owen, M. & Yoshida, R. 2009. First steps toward the geometry of cophylogeny. arXiv preprint 0809.1908v03

Jayawardhana, B., Kell, D.B. & Rattray, M. 2008. Bayesian inference of the sites of perturbations in metabolic pathways via Markov chain Monte Carlo. *Bioinformatics 24(9): 1191–1197.*

Jensen, L.J., Julien, P., Kuhn, M., von Mering, C., Muller, J., Doerks, T. & Bork, P. 2008. eggNOG: automated construction and annotation of orthologous groups of genes. *Nucleic Acids Res. 36: D250–D254.*

Junga, S., Lee, K.H. & Lee, D. 2007. H-CORE: Enabling genome-scale Bayesian analysis of biological systems without prior knowledge. *Biosystems 90(1): 197–210.*

Klau, G.W. 2009. A new graph-based method for pairwise global network alignment. *BMC Bioinformatics 10(Suppl 1): S59.*

Koch, I., Junker, B.H., Heiner, M. 2005. Application of Petri net theory for modelling and validation of the sucrose breakdown pathway in the potato tuber. *Bioinformatics 21(7): 1219–1226.*

Koonin, E.V. & Wolf, Y.I. 2008. Genomics of bacteria and archaea: the emerging dynamic view of the prokaryotic world. *Nucleic Acids Res. 36(21): 6688–6719.*

Koski, T. & Noble, J. 2009. Bayesian Networks: An Introduction. *Wiley series in probability and statistics.*

Kuzniar, A., van Ham, R.C.H.J., Pongor, S. & Leunissen, J.A.M. 2008. The quest for orthologs: finding the corresponding gene across genomes. *Trends in Genetics 24(11): 539–551.*

Lähdesmäki, H., Rust, A.G., Shmulevich, I. 2008. Probabilistic inference of transcription factor binding from multiple data sources. *PLoS ONE. 3(3): e1820.*

Latino, D.A., Zhang, Q.Y. & Aires-de-Sousa, J. 2008. Genome-scale classification of metabolic reactions and assignment of EC numbers with self-organizing maps. *Bioinformatics 24(19): 2236–2244.*

Lee, P.H. & Lee, D. 2005. Modularized learning of genetic interaction networks from biological annotations and mRNA expression data. *Bioinformatics 21: 2739–2747.*

Lemmens, K., Dhollander, T., De Bie, T., Monsieurs, P., Engelen, K., Smets, B., Winderickx, J., De Moor, B. & Marchal, K. 2006. Inferring transcriptional module networks from ChIP-chip-, motif- and microarray data. *Genome Biology 7(5): R37.*

Le Novère, N. *et al.*, 2009. The Systems Biology Graphical Notation. *Nat Biotechnol. 27(8): 735–41.*

Loewenstein, Y., Portugaly, E., Fromer, M. & Linial, M. 2008. Efficient algorithms for accurate hierarchical clustering of huge datasets: tackling the entire protein space. *Bioinformatics 24: i41–i49.*

Lukashin, A.V. & Borodovsky, M. 1998. GeneMark.hmm: new solutions for gene finding. *Nucleic Acids Res 26: 1107–1115.*

Majoros, W.H., Pertea, M. & Salzberg, S.L. 2004. TigrScan and Glimmer-HMM: two open source ab initio eukaryotic gene-finders. *Bioinformatics 20: 2878–2879.*

Marcet-Houben, M. & Gabaldón T. 2009. The tree versus the forest: the fungal tree of life and the topological diversity within the yeast phylome. *PLoS ONE. 4(2): e4357.*

Markowetz, F. & Spang, R. 2007. Inferring cellular networks - a review. *BMC Bioinformatics. 8(Suppl 6): S5.*

Paten, B., Herero, J., Beal, K., Birney, E. 2009. Sequence progressive alignment, a framework for practical large-scale probabilistic consistency alignment. *Bioinformatics. 25(3): 295–301.*

Rentzsch, R., Orengo, C.A. 2009. Protein function prediction - the power of multiplicity. *Trends Biotechnol. 27(4): 210–9.*

Ridder, L., Wagener, M. 2008. SyGMa: combining expert knowledge and empirical scoring in the prediction of metabolites. *Chem Med Chem 3(5): 821–832.*

Salamov, A.A. & Solovyev, V.V. 2000. Ab initio gene finding in Drosophila genomic DNA. *Genome Res 10: 516–522.*

Segal, E., Shapira, M., Regev, A. *et al.* 2003. Module networks: identifying regulatory modules and their condition-specific regulators from gene expression data. *Nature Genetics 34: 166–176.*

Semple, C. 2008. Hybridization networks, in: O. Gascuel, M. Steel (eds.), Reconstructing evolution: New mathematical and computational advances, *Oxford University Press, pp. 277–314.*

Sen, I., Verdicchio, M.P., Jung, S., Trevino, R., Bittner, M. *et al.* 2009. Context-specific gene regulations in cancer gene expression data. *Pac Symp Biocomput. pp. 75–86.*

Shih, A.C., Lee, D.T., Peng, C.L., Wu, Y.W. 2007. Phylo-mLogo: an interactive and hierarchical multiple-logo visualization tool for alignment of many sequences. *BMC Bioinformatics 8: 63.*

Tanay, A., Sharan, R. & Shamir, R. 2002. Discovering statistically significant biclusters in gene expression data. *Bioinformatics 18 Suppl 1: S136–S144.*

Tatusov, R.L., Fedorova, N.D., Jackson, J.D., Jacobs, A.R., Kiryutin, B., Koonin, E.V., Krylov, D.M., Mazumder, R., Mekhedov, S.L., Nikolskaya, A.N., Rao, B.S., Smirnov, S., Sverdlov, A.V., Vasudevan, S., Wolf, Y.I., Yin, J.J. & Natale, D.A. 2003. The COG database: an updated version includes eukaryotes. *BMC Bioinformatics 4: 41.*

Van den Bulcke, T., Lemmens, K., Van de Peer, Y. & Marchal, K. 2006. Inferring Transcriptional Networks by Mining Omics Data. *Current Bioinformatics, 1(3): 301–313.*

Virtanen, H.E, 1995. A Study in Fuzzy Petri Nets and the Relationship to Fuzzy Logic Programming. *Turku Centre for Computer Science, reports, A95–162, 30 pp.*

Vizcaíno, J.A., Mueller, M., Hermjakob, H. & Martens, L. 2009. Charting online OMICS resources: A navigational chart for clinical researchers. *Proteomics 3(1): 18–29.*

Wang, W., Cherry, J.M., Botstein, D. & Li, H. 2002. A systematic approach to reconstructing transcription networks in Saccharomyces cerevisiae. *Proc Natl Acad Sci U S A 99: 16893–16898.*

Chapter 9

Future Challenges

In modern post-genomics and systems biology, the increasing flood of data gives heterogeneous, partial and overlapping information about gene variations, gene and protein expression, metabolic events and connections in metabolic and regulatory networks. The challenging task is to screen out the unreliable pieces of data and to connect and integrate the multitude of data with connections and subgraphs of various bionetworks to a coherent whole which can be validated and compared with other datasets with better annotation. This chapter signposts recent trends and developments, with examples, in applying AI methods in network analysis, unsupervised clustering and supervised learning (classification), as well as neural network and evolutionary methods (Genetic Algorithms). The rise of ontologies to the challenge of metadata harmonization and use of ontologies as data is covered in the penultimate subchapter. In the final subchapter the computational bottleneck and the use of GPUs and cloud computing and self-indexes for massive datasets are discussed.

9.1 Network analysis methods

Algorithms for graph comparisons and network matching are important methods in telecommunications, internet networking, computer chip design and so on, and such techniques are already being usefully adapted to bioinformatics problems in the systems biology domain. However, in biology such data analysis tasks are often more challenging than in engineering, because the data is more heterogeneous and collected using a wide range of experimental or *in silico* methods with varying accuracies and reliabilities. Often even harmonizing or pooling the data labeling and metadata between different data sources poses problems at the outset of the analysis.

A general trend in network analysis methods is that new formulations for aligning sparse networks keep appearing, and hybrid methods involving graph matching and other techniques are being introduced, in addition to traditional Bayesian network analysis. Network analysis methods for large matrices is an established field in other disciplines (physics, engineering, data mining) and include various random walk and diffusion methods for graph clustering and dimensionality reduction (*e.g.* Nadler *et al.* 2005, Skillikorn 2007). Some of these techniques are related to the factorization methods and self-indexes discussed in other chapters of this book; for the latest paper on short codes related to spectral hashing, see Jégou *et al.* (2009). Currently normalized spectral clustering methods are considered most consistent and assumption-free (von Luxburg *et al.* 2008). Such advanced spectral clustering methods for graphs and networks could be used more in bioinformatics as well.

An example of such novel algorithms for alignment of sparse networks (which is an NP-hard problem) is that of Bayati *et al.* (2009), who utilized belief propagation among near neighbours in the network and distributed processing. It showed very good performance compared to conventional global graph matching methods, even on large networks with over 100,000 vertices in the network, and it appears that the approximated solutions are within 95% of optimum solutions. Global network alignment, on the other hand, is also improving and has recently been formulated as an eigenvalue problem of the network describing matrices. Such a formulation and was used by Singh *et al.* (2008) for identification of evolutionarily related (orthologous) genes.

Hybrid methods mixing various techniques with network analysis are also becoming more important. An elegant example how metabolic network data can be used is the recent elucidation of coevolution between enzyme activities and the genes coding for them by Schütte *et al.* (2009), who showed a correlation between KEGG pathway node distances of enzymes and the corresponding sequence similarities. Such analysis would not be possible without the systems biology type comprehensive metabolic network data already accumulated in the KEGG pathway database and other similar databases. Another interesting example is predicting nutritional requirements by SVM using metabolic network properties of the organism from KEGG pathway database (Tamura *et al.* 2009).

Similarly, a protein interaction database data was used by Yao & Rzhetsky (2008) to predict the probability of a protein/gene being a good drug target by a Bayesian network, logistic regression and Naïve Bayes methods. They utilized a set of 919 FDA approved drugs and network data from BIND database of interacting proteins, known metabolic pathways and BRENDA tissue ontology. The integrated use of the different data sources led to reliable quantitative prediction of gene druggability, based largely on network connectivity and tissue expression profile of the target gene.

In conclusion, the existing biological network databases are already a very valuable data source, which will grow in importance, as large-scale systems biology data collection projects progress. Therefore bioinformaticians need to learn how to integrate complicated network data with other traditional types of bioinformatics data for efficient knowledge and hypothesis generation about functions and interactions of genes, proteins and metabolites. For further ideas on available methods, the review of Alves *et al.* (2008) is recommended.

9.2 Unsupervised and supervised clustering

For unsupervised learning (clustering without a validation data set) a very large variety of approaches is used, including k-nearest neighbours, CART, and so on, as discussed by Bickel (2009); he considers currently random forests and boosting methods the most effective in general. Many good modern statistical learning methods suitable for bioinformatics applications are covered in a book by Hastie *et al.* (2009). A useful review of current clustering methods for biomedical applications is in Andreopoulos *et al.* (2009), which compares over 40 different methods, including also the promising non-negative matrix factorizations, as discussed already elsewhere in this book.

Supervised clustering by support vector machines and kernel development for new types of bioinformatics data are currently in the forefront of recent progress in analyzing systems biology scale postgenomic data. SVM is a powerful method, provided a kernel or a combination of kernels well suited to the specific type of biodata being analysed can be devised. Kernels can be effectively combined and optimized by semi-definite programming, as explained in Lanckriet *et al.* (2004), but such combination kernels have not yet been used extensively and are worth more attention.

SVM can also be combined with a variety of other techniques for preprocessing or postprocessing or as part of an iterative solution finding by a data space exploring algorithm and solution candidate evaluation by SVM. Recently Onuki *et al.* (2009) successfully combined Hidden Markov Model with SVM evaluation of solution candidates to analyse genetic haplotypes to predict phenotype. It is expected that hybrid methods including SVM as one module will be in the mainstream of many future bioinformatics methods for complex high-dimensionality data.

The challenge for both unsupervised clustering and supervised classification hybrid methods is to evaluate, validate and scale up these methods so that they can utilize the large complex datasets becoming commonplace in bioinformaticians' hands. Hybrid methods with efficient data space exploration and incorporating

SVM in one form or another seem to be the most promising in this respect.

9.3 Neural networks and evolutionary methods

In spite of the success of the SVM methods, the neural networks approach is still very popular and has been shown to perform well in many cases on biological data. Neural network methods have been criticized by being a kind of arbitrary black box and involving unrealistic modeling of the interactions of the real biological variables. However, the success of neural network methods in classification of large microarray and mass spectrometry data is demonstrated in the review of Lancashire *et al.* (2009). In analyzing that kind of data, the success of classification matters more than the actual biological reality of the model.

As an example, modern neural network models like GPNN and GENN are considered promising for gene selection problems in genetic epidemiology (Motsinger-Reif & Ritchie 2008). A further interesting observation is that switching neural network with recursive feature addition (SNN-RFA) beats SVM with recursive feature elimination (SVM-RFE) in gene selection from microarray data (Muselli *et al.* 2009). Similarily, time lagged recurrent neural network are claimed to be better than SVM or SOM analysis (Liang & Kelement 2009) for gene expression analysis. Neural networks have also been used successfully in bionetwork interaction discovery, *e.g.* by Lemetre *et al.* (2009).

The final jury is still out on whether neural networks will be competitive in the long run, as SVM methods continue to improve and even evolutionary algorithms (discussed below) are becoming more attractive due to their better searching abilities in complex data spaces and to their being ported to computationally powerful cloud computing systems and GPU farms (see last subchapter).

Evolutionary algorithms are currently increasing in popularity and thus competing with SVM for attention among bioinformaticians. These genetic algorithms emulate natural evolution by generating iteratively populations of solutions from which best ones are selected for making the next improved solution candidate. Such methods have been considered an *ad hoc* method of searching large data spaces without much biological reality in the model itself. However, they have been found to give good solutions, although the optimum solution is not guaranteed theoretically. The Genetic Algorithm type methods can explore efficiently complex data spaces, but have the disadvantages of complex setting up, high computational demands and no overarching theoretical basis on what features to include in the feature set and how to evolve and select the candidate solutions, i.e. how to set up the fitness function. However, they can easily be configured for multi-objective optimization and are also easily parallelized, which is a good feature for GPU implementation (see last subchapter on GPUs).

Thus evolutionary algorithms suit problems with complex data spaces and with lots of available computing resources, when the interpretability of the model is not so important. A recent review (of a proponent of Genetic Algorithms) concludes that such methods still have their place among pattern recognition methods (Orriols-Puig *et al.* 2008), and many bioinformatics problems can to be attacked by evolutionary methods.

As mentioned before for neural network methods, combining SVM with a Genetic Algorithm into a hybrid method can boost performance of classification, which has already been demonstrated in a several cases for microarray gene selection (Duval & Hao 2009, Dess & Pes 2009). Whether this will be the case for other types of biodata classification problems remains to be seen.

Some matrix sorting formulations of biological complex network data can be formulated as the quadratic assignment problem (QAP) and such a method has been used to analyse cancer microarray data (Tsafrir *et al.* 2005). However, such calculations are very computing intensive, but this could be alleviated by sampling the possible permutation solutions by a parallelized Genetic Algorithm implemented in GPU platform. GPU implementation of QAP was recently developed by Tsutsui & Fujimoto (2009). In general it seems that Genetic Algorithms could benefit from GPU implementations, which could make these methods faster and thus more popular.

Estimation of distribution algorithms (EDAs) are a new promising alternative to Genetic Algorithms (Armananzas *et al.* 2008). EDAs have the advantage over genetic algorithms in not having many parameters which have to be set up and tuned. Genetic algorithms also need a lot of work to set up correctly the required population increases, crossing-over of features and selection intensity parameters. Further advantages of EDAs are that previous knowledge about the problem can be included in the starting model, and that in the middle of the analysis process the new emerging patterns or interactions between variables can be accessed from the generated models during their improvement. EDA seem to be most useful in high dimensionality and NP-hard problems. The general workflow of an EDA is shown in Figure 9-1.

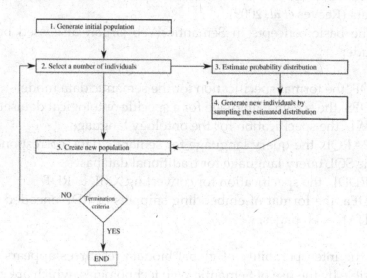

Figure 9-1. EDA principle. From Armananzas *et al.* 2008. Reproduced by BioMed Central Open Access Charter.

Even more interesting from a methodological point of view are the quantum computing inspired methods, which are starting to appear in bioinformatics literature, too. A multimodel EDA framework was used in a novel quantum-inspired evolutionary algorithm (Platel *et al.* 2010) with good exploration of complex data space and fast convergence. Similarly, Xiao *et al.* (2009) used an evolving hybrid quantum chaotic swarm to encode DNA efficiently and Abderrahim *et al.* (2009) used a hybrid genetic and quantum algorithm for gene selection from microarray data. Quantum-inspired methods may invigorate evolutionary algorithms to make them more attractive due to better computational efficiency and better finding of optimum solution.

9.4 Semantic web and ontologization of biology

Huge amounts of genome and proteome annotation keep accumulating, as discussed in the beginning of this book. This poses enormous challenges in harmonizing and pooling data from different sources due the multitude of formats and categorization of data (Reeves *et al.* 2009).

The basic concepts in Semantic Web jargon discussed below include:

• RDF, the format specification for the semantic data model
• RDFS, the schema language for a specific ontological dataset
• OWL, the specification for the ontology language
• SPARQL, the query language for semantic data, corresponding to the SQL query language for traditional databases
• GRDDL, the specification for converting XML to RDF
• RDFa, the format of embedding snippets of RDF encoded data in HTML web pages.

True interoperability of global biodata resources appears only possible by the use of Semantic Web technologies, which are being developed to Linked Data On Web (LDOW) services to enable machine queries of ontologized data (Berners-Lee 2009). A large network of biological data is already linked this way, including PubMed, UniProt, KEGG, Gene Ontology and so on, and already contains billions of indexed RDF/OWL statement triplets. Many of the already ontologized biodata sources have set up Linked Data services in the LDOW network. NCBI is already setting up a BioPortal to pool Open Biomedical Ontologies and other RDF/OWL ontologies in one portal website to give a single access point to such data (Noy *et al.* 2009).

The Linked Data concept is an advance over the earlier popular formatting of concepts and controlled vocabularies into to XML for automated computer search and display. In XML only simple hierarchical relationships are possible, but in Linked Data on the Semantic Web a large variety of information can be encoded in triplets of the form *subject => relationship => object* in RDF format. A network of such triplets forms a network of related concepts and then serves as placeholder for data items matching the relationship of the RDF triplets. Linked Data web services then link the RDF triple databases through HTTP protocol on the internet, with embedded links to ontologically defined data resources. All data is encoded into triplets, and the huge repositories of triplets can be queried by the special query language called SPARQL.

The aim of the ontologization of biodata into RDF triples is to allow databases to be accessed by computer algorithms based on semantic concepts rather than by the specific query fields given by humans or SQL statements. This would allow to use ontologies as data, so that relatedness of concepts linked to data items would be a classification variable just like any other data item. Similarity by semantic distance promises to enhance the accuracy of searching and classifying relevant items compared to the old style word-based Boolean searches.

A big challenge is still developing easy to use tools for finding and utilizing the Linked Data services, because there is a multitude of ontologies of different complexities pertaining to a large variety of bioinformatics data (as well as other types of data). Also, mechanisms are needed for researchers and users of the data to publish small RDF triple datasets without a heavy and complex ontology infrastructure.

One mechanism to publish and disseminate such RDF triples is RDFa annotation of HTML web pages, which has already been used in NCBI Entrez system for neuroscience research (Samwald *et al.* 2009). Semantic web query systems using SPARQL have already been implemented for such a repository of data for neurosciences (Cheung *et al.* 2009). Such nanopublication of biological facts and experimental results in atomized snippets of information. Such a mechanism could make it easy for anybody to create standardized biological knowledge via web publishing.

One suggested improvement to RDFa tagging is adding to published web pages associative tags (aTags) using snippets of XHMTL/RDFa embedded in RDF/OWL, thus making accessible individual factual statements one by one, embedded in web pages for semantic search engines to discover (Samwald & Stenzhorn 2009). The tags would conform to Open Biomedical Ontologies and would be findable and queriable by a semantic web crawler.

This very interesting concept of nanopublication as a mechanism of scientific discourse is discussed further in Mons & Velterop (2009). They conclude that this could be a new way of valuing scientific contributions and an easy mechanism to accumulate important biological facts as snippets of *e.g.* RDF triples. Then the mechanisms of actually utilizing and comparing conflicting and corroborating nanopublications will become the next challenge, involving trust-worthiness of publisher, resolution of conflicting statements and the value/crediting of repeating the same facts/experiments by different people as well as the remaining role of traditional journal articles.

One solution for the trust, evidential and provenance information is the use of named RDF graphs, in which a sub-network of connected RDF triples is named, given a web reference (URI), author name, date and so on (Zhao *et al.* 2009a). Each data link is also referenced by its' reason of establishment, update time, history of changes and so on. Such a system has already been established in the FlyWeb, a semanticized web resource for fruitfly researchers. Another interesting example, highlighting also the problems in matching ontologies and overlapping distinct concepts in related disciplines, is the integration of ontologies and data for Western and traditional Chinese medicine (Chen *et al.* 2007, Zhao *et al.* 2009b).

Having all these data available as RDF triples, however, is only the first step. Easy to use tools are needed to search and compute with such resources, *e.g.* easy ways to design suitable SPARQL queries and limitation of the extent of web crawling when the search agents are gathering relevant data. A wide front effort is needed to develop both interoperable ontologies and fast methods of utilizing ontologized data by novel algorithms that can handle huge datasets of biotechnical and chemical research as well as related intellectual property (Havukkala 2009). A further tricky issue is matching data from ontologies with overlapping or conflicting concepts.

The first generic semantic search engines for web searching are already appearing on the internet, and some biology specific semantic web searches have been established also. The GoWeb semantic web search for life sciences (Dietze & Schroeder 2009) ranks and filters snippets of data hits by relevant Gene Ontology categories and MeSH medical concept annotation (*i.e.* by conceptual distance to the target of the query). The system produced improved results compared to traditional state of art systems for gene function analysis, diseases and symptoms identification and linkages between proteins and diseases. GoWeb represents a proof of concept for semantic search technologies for bioinformatics. A similar integrated semantic search system for gene expression data reduced the time to find the answers to almost half compared to manual browsing of a small set of gene expression databases by users (Sutherland *et al.* 2009).

Semantic technologies are starting to become mainstream also in bioinformatics, but it will take quite some time before the old web services are updated and migrated to the Semantic Web format and end users are provided with the necessary browsing·

and searching tools for dataset gathering and subsequent data analysis. Here the challenge of data fusion from heterogeneous sources is formidable and is discussed next.

As for text mining in the Semantic Web context, few attempts have yet been made to integrate semantic information with word sense disambiguation. Recently, semantic similarity of the co-occurring words was used to improve assignment of correct meanings to ambiguous Gene Ontology terms (Alexopoulou *et al.* 2009). This is a first step in automated context-aware data mining starting from free text documents and taking advantage of domain-specific ontologies. Generic free text mining for semantically correct information from scientific biology publications is, however, still far in the future.

9.5 Biological data fusion

The machine learning community has already long practiced data fusion by integration of multimodal data for enhanced classification, and such methods are starting to appear in bioinformatics, too. As an example, Lähdesmäki *et al.* (2008) used probabilistic models containing multimodal data for transcription factor binding site prediction and obtained better results than using sequential classification using the same data sources. Liu *et al.* (2009a) fused data from over 3000 MESH etiology related headings with UMLS semantic network and defined 1100 genes associated with diseases, obtaining a good joint clustering of diseases, genes and etiologies.

In addition to data fusion in classification (supervised clustering), data fusion can be also usefully implemented in unsupervised clustering. The main methods of data fusion are discussed in Yu *et al.* (2009). The traditional ensemble clustering methods use two or more clustering algorithms, and join the partitions of data from each algorithm using a consensus function. In the data fusion approach, Yu *et al.* (2009) distinguish two main approaches: 1) using one clustering algorithm and joining two partitions of data from each algorithm using a consensus function; and 2) using a combination of two of more different kernel matrices in SVM analysis with the clustering algorithm to obtain partitioning of data by data fusion. Figure 9-2 shows the flow diagrams of these approaches.

Extensive experiments (including also bioinformatics datasets) by Yu *et al.* (2009) showed that data fusion is effective and improves clustering results and that kernel fusion in general is more effective for heterogeneous data than ensemble clustering with data fusion. Such methods are thus recommended for biodata analysis, when possible.

Ideally data fusion includes incorporation of semantic data as well, which then can be integrated with machine learning methods for clustering and classification (Kasabov *et al.* 2008). Ultimately, such ontologized data analysis in an integrated fashion will aid biomedical decision support for global, local and personalized models of diseases and gene interactions analysis.

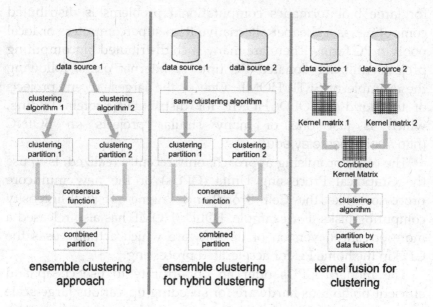

Figure 9-2. Data fusion frameworks. From Yu *et al.* (2009). Reproduced by permission from the authors.

9.6 Rise of the GPU machines

The famous Moore's law states that the density of transistors on a chip doubles every two years, and this trend has held since 1960s. However, it is now predicted that the commercially feasible minimum feature size is around 20 nanometers, and that the end of Moore's law is going to happen around 2015. This will mean that the performance of CPUs and high-density memories will soon reach a ceiling. Current densest memories store around 40 electrons per cell and going to a lower count will decrease reliability of these devices. Similarly, physically feasible size limits are being approached for CPUs. Development of the next-generation of 3-dimensional memory and CPU structures is far in the future and is not likely to be in time to alleviate the looming performance gap for processing power for the exponentially growing data glut in bioinformatics, especially from high-throughput and personalized genomics and systems biology.

One way to tackle the problem of insufficient CPU cycles for large bioinformatics computational problems is distributed computing, as a cheaper alternative than supercomputers or local pools of PC farms. There are many such distributed biocomputing projects utilizing donated PC time of internet users, following the example of SETI@HOME. One of the largest recent projects of this kind is FOLD@HOME for distributed protein folding, which is just one of many similar projects on BOINC (http://boinc.berkeley.edu/).

The other promising approach, only recently realized, is to use the Graphical Processing Units (GPUs) or the new multicore processors like the Cell Broadband Engine for high-intensity computing tasks. For example, FOLD@HOME has also released a more advanced version of its software which actually uses the GPU in the home PC for accelerated processing.

The modern GPUs offer an opportunity to use cheap and efficient ubiquitous hardware for speeding up various large-scale computations, including bioinformatics. It, however, necessitates that the programming technology can be made standardized and easy enough to use for bioinformaticians, who are not normally well versed with low-level programming languages. In other fields GPU computational research is already well established. A notable recent programming platform for such computing is NVIDIA's CUDA programming environment for NVIDIA GPUs (see Figure 9-3) as well as the Jacket module in MatLab to access GPU resources directly from MatLab routines. Recent efforts of using GPUs in biocomputing are highlighted here to give a taste of this very new and promising field in algorithmic bioinformatics.

Modern traditional CPUs can process large number of parallel instructions in the mode of Multiple Data Multiple Programs (MDMP). At the other end of the spectrum are GPUs, which contain a large number of small processors which perform Single Instruction Multiple Data (SIMD) in parallel, for repetitive tasks like graphics image processing, ray tracing *etc*. In a very specialized category are the even faster programmable FPGAs (Field-Programmable Gate Arrays) which may perform well for some specialised tasks in bioinformatics.

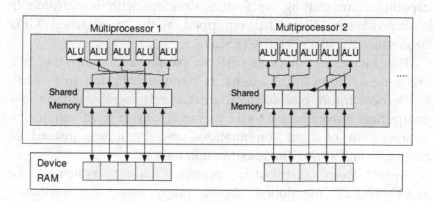

Figure 9-3. The CUDA compatible GPU architecture. The arrows between ALUs (Arithmetic Logic Units) and Shared Memory show that Multiple Instructions for Multiple Data) MIMD is possible. From Manavski & Valle 2008. Reproduced by Creative Commons Attribution License from an Open Access article.

FPGAs have been set up to accelerate Smith-Waterman similarity search (Karanam *et al. 2008*), but general interest has moved to much cheaper and fast improving general purpose GPUs. FPGA may still have a role in very specific niche computation, but are not discussed further here.

The IBM Cell BroadBand Engine (CBE) used in PlayStation 3 is an intermediate technology between pure CPU computation and GPU computing, in which each of the eight processing units is more or less independent, giving it both faster processing speed than standard CPUs and more flexible data use than GPUs. In terms of processing power and flexibility, the new multicore chips like CBE gain the best features of both CPUs and GPUs, but with the cost of more difficult programming, because the algorithms need to be parallelized and only small amounts of fast memory is available in each subprocessor.

A CBE has 8 processing units, each with 25 GFLOPs, thus having a maximum compute power of 200 GFLOPS, compared with a standard modern two-core CPUs having 15-20 GFLOPS and high-end GPUs (*e.g.* NVIDIA G8 series has 128 processors) generating over 500 GFLOPS. It appears that supercomputer capabilities are coming to scientists' desktops with new affordable multiprocessors like CBE equipped with an attached GPU processor card like NVIDIA TESLA.

This increased power to perform parallelized computing is a welcome addition to the arsenal of bioinformaticians. In essence, in the near future bioinformaticians can rely increasingly on raw computing power to use exact methods, which need exhaustive enumeration of data combinations and iterations, instead of heuristics and approximate estimation methods.

Apart from distributed protein folding systems like FOLD@HOME mentioned above, many other bioinformatics problems also need huge computing power for an exhaustive all against all comparison of data items. Such parallelizable tasks include assembling genome fragments to clusters of overlapping sequences, finding related sequences for clustering in a metagenomic dataset for simultaneous assembly of several genomes from a mixed sample, detecting contaminations from other species or sequencing vectors in a genome assembly and microarray data analysis, to mention a few.

While many improvements in speed were gained by moving the exact exhaustive sequence search algorithm Smith-Waterman (which guarantees an optimal alignment result) to heuristic faster methods like BLAST and BLAT, more speedup is still necessary, due to the exponential increase in genomic datasets. Already comparisons of all known genomes to each other are attempted for large-scale evolution studies. In 2008 a project to sequence 1000 human genomes was started and these genomes will need to be compared to each other. Therefore superior genome alignment speed is mandatory to handle the new large datasets.

An example of this kind of CPU based software is ScalaBlast (Oehmen *et al.* 2006), which is order of magnitude faster than BLAST and has already been used to compare over 30 bacterial genomes to each other by performing 2 billion pairwise sequence comparisons (Shah *et al.* 2007). Another novel system called SPALN for high-throughput alignment of a set of protein sequences (proteome) or a set of ESTs to a genome was developed by Gotoh (2008a, 2008b) and runs up to two orders of magnitude faster than traditional BLAST or 10 times faster than MegaBlast similarity search on standard CPU systems. SPALN is based on looking for elevated frequencies of short 10–12 k-mers from the query sequence in blocks of genomic sequence to identify the approximate location of a match and then doing a further search with the identified block of sequence.

But why use ScalaBLAST or SPALN on a CPU based systems, if you could instead use an even faster GPU implementation of Smith-Waterman search or BLAST/BLAT, or even better, a GPU accelerated version of ScalaBlast or SPALN?

In fact, such systems are already appearing. A Smith-Waterman algorithm has been published for the Playstation CellBE (Wirawan 2008) and another implementation achieved a 55-fold speedup compared to high-end multicore CPU computation (Aji & Feng 2008a, 2008b). Using a CBE based personal computer, Smith-Waterman can be run at the same speed as BLAST (Swalkowski *et al.* 2008, Manavski & Valle 2008) and was already used to align ESTs to genome sequence (Manavski *et al.* 2008). Multiple GPU parallelization of Smith-Waterman has also already been shown to be very efficient (Jung 2009). The promise and challenges of doing Smith-Waterman searches with GPUs is reviewed by Striemer & Akoglu (2009) and Liu *et al.* (2009b) already reported a GPU implementation which is 3–10 times faster than standard NCBI Blast.

BLAST has been accelerated up to 20 times by GPU implementation to utilize an order of magnitude larger datasets (Nguyen & Lavenier 2008), and BLASTP was accelerated by GPU by Zhang *et al.* (2008). Many other such implementations are being developed around the world.

GPU processing is likely to be also very useful for many Hidden Markov Model algorithms. As a proof of principle, a large speech recognition HMM model was accelerated almost ten-fold by NVIDIA G-80 GPU card compared to standard CPU (Chong *et al.* 2008). Similarly, neural network classification of images was 5-fold faster using GPU than CPU, and 10 times faster using an integrated solution (Jang *et al.* 2008). The company Scalable Informatics has in early 2009 released a GPU-HMMER implementation for NVIDIA GPUs for protein homology analysis (Anon 2009). MEME motif finder has also been implemented for GPU with one order of magnitude speedup with one GPU processor and two orders of magnitude speedup with a cluster of GPUs (Chen *et al.* 2008) compared to standard CPU processing.

Some novel algorithms for string matching designed for standard processing in CPUs are already showing orders of magnitude faster performance compared to old algorithms. Many of these are expected to become even faster, if they can be ported to GPUs. An example is high-throughput matching of short sequence reads to large genomes, which is one of the most important applications in the current era of ultrahigh-throughput sequencing. The most recent CPU software acceleration for this task, BOWTIE (Langmead *et al.* 2009a), is a fast implementation of

Burrows–Wheeler transform (see Figure 9-4) and can align about 30 million sequence reads per hour on a standard PC with 2 gigabytes of RAM. This rate is 60 times faster than the current state of art CPU based algorithm MAQ (Li *et al.* 2008b, http://maq.sf.net) on a PC and over 350 time faster on a server machine than a similar SOAP algorithm (Li *et al.* 2008a).

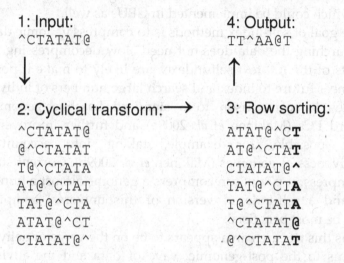

1: Input:
^CTATAT@

4: Output:
TT^AA@T

2: Cyclical transform: ⟶

^CTATAT@
@^CTATAT
T@^CTATA
AT@^CTAT
TAT@^CTA
ATAT@^CT
CTATAT@^

3: Row sorting:

ATAT@^CT
AT@^CTAT
CTATAT@^
TAT@^CTA
T@^CTATA
^CTATAT@
@^CTATAT

Figure 9-4. The Burrows–Wheeler transform creates more easily compressible sequences and an index on the long target sequences enables faster comparison of a short string to the long target. The original string (1) is rotated one character at a time (2), then the rows are sorted alphabetically (3), and the last column taken for the output string (4), which contains more successive repeated characters (TT and AA) that compress better. The backward transformation is possible to obtain the original string.

BOWTIE has already been used as an embedded module in assembling short messenger RNA reads to a genome in TOPHAT algorithm with good results (Trapnell *et al.* 2009). For more possible applications of Burrows–Wheeler transform for biodata, see the recent book by Adjeroh *et al.* (2008).

Another dramatic speed-up using a different method of k-mer processing was developed by Arnau *et al.* (2008), using oligomers of length 13 (which is usefully a prime number) to search for matches in a long genome sequence. A similar method for repeat enumeration in genomes was published by Kurtz *et al.* 2008).

All these new k-mer based clever methods are related to compressed full-text self-indexes (Ferragina *et al.* 2009a), which are maturing from theory to practical applications for previously unparalleled speed and small memory space. Theoretical advances have been made for motif finding (Nicolas *et al.* 2009) as well, which could be implemented in GPUs as well.

One goal of self-index methods is to compress genomic data so that searching the data does not need slow decompressing. Such variants of the full-text self-indexes are likely to make it possible in the near future to index and search large numbers of individual genomes by comparison to a standard baseline genome on standard PCs (Mäkinen *et al.* 2009a) and further improvements appear possible, for example, taking into account the repetitiveness of genomes (Mäkinen *et al.* 2009b). Current state of art compressor, p7zip, can compress a genome almost a hundred-fold, and a self-indexed version of this universal compressor would be most beneficial.

Thus this research area appears to be on the verge of delivering solutions to the post-genomic wave of data and the envisaged distributed processing. Siren (2009) has very recently developed an efficient Compressed Suffix Array (CSA) indexing method, which is parallelizable for distributed computing and can index several gigabytes an hour running on a 16-core linux system with four quad-core Xeon X7350 processors. Such a system can index the complete human genome in about 2 hours with only 3 gigabytes of memory. Another recent algorithm by Ferragina *et al.* (2009b) obtains fast light-weight indexing by sequential access to disk using Burrows–Wheeler Transform (BWT).

If these algorithms can be implemented to take advantage of the GPU processing power, further orders of magnitude speedups could be achievable, especially for the latest k-mer and suffix tree based methods discussed above.

An important reason for the attractiveness of k-mer methods, apart from being inherently suitable for GPU acceleration, is that modern DNA sequencing machines output massive amounts of 30–80 basebairs long sequence reads. These short DNA reads can be handled as long k-mers for fast assembly of genomes or for search of matches to target mutations, identification of pathogens and so on. Not only can large amounts of randomly selected sequences be sequenced in brute force to assemble full genes or genomes at will, but also targeted massively parallel sequencing of *e.g.* all 20,000 protein coding segments showing differences between human individuals from the "standard" genome is becoming possible. The newest experimental method uses 170 bases long oligonucleotides for capturing targets and sequencing them by 76 basepairs long reads by Illumina GA-II machines (Gnirke *et al.* 2009). Other similar methods will increase even further the data flood for bioinformaticians to process. Schmidt *et al.* (2009) have already developed GPU based fast short read assembly tools, and more tools like this are expected to be available soon.

One further area of great potential for GPU use in bioinformatics is the speeding up of microarray analysis. In principle, all neural network methods which analyze large numbers of features would benefit, including gene selection problems for disease-prediction using microarray gene expression data. For example, k-means and KNN algorithms have already been implemented for CBE (Buehrer *et al.* 2008). They used GPU processing in semi-supervised classification methods involving iterative cross-validation analysis by test and validation data samples. The result was an efficient systems biology graph network analysis to select the network that best matched the data. A genetic programming algorithm adapted for GPU processing was recently used to analyze large amounts of gene expression data (Langdon & Harrison 2008, Langdon 2008). Similarly, Chang *et al.* (2008) used a GPU implementation for agglomerative hierarchical clustering of microarray data to get 40-fold acceleration over standard CPU processing, because calculation of pairwise Euclidean distances could be done much more efficiently in parallel in the GPU processors. Other methods useful for microarray analyses have been accelerated by GPUs as well. For example, Ma & Agrawal (2009) developed a high level programming interface and developed GPU code for k-means clustering, EM clustering and Principal Component Analysis (PCA) which performed 20–50 times faster than original CPU based code.

Phylogenetics problems also involve searching through large numbers of alternative phylogenetic tree topologies to find the most likely sequence of evolutionary events matching the observed data. With GPU systems it will be also possible to simulate large numbers of possible scenarios of genome rearrangements to arrive at the most likely history of DNA translocations observed within one species or between related species. Genetic algorithms may become important in this field to explore the large data space efficiently, where even GPUs cannot do exhaustive brute force analysis. For a review of genetic algorithms already ported to GPUs, see Banzhaf *et al.* (2009). Many of these could be used for bioinformatics problems as well.

GPU acceleration is also needed for future analysis of large gene expression datasets based on brute force sequencing of cDNA tags instead of analog microarrays. In brute force sequencing the number of counts for sequenced cDNAs corresponds to the hybridization intensity. Such data demands large amounts of fast alignments and comparisons of sequence tags to identify the expressed genes reliably. A quantitative comparison between microarray and short read sequencing technologies (Bloom *et al.* 2009) showed that microarrays are still better in finding differential expression for low-abundance mRNAs, but sequencing can detect alleles and splice variants more reliably. It is expected that future high throughput sequencing systems of this kind will come with GPU-based accelerated sequence matching and alignment software.

Such software and new algorithms suitable for GPU are already emerging. Kloetzli *et al.* (2008) developed a CPU + GPU optimised algorithm for finding the longest common subsequence from very long target sequence with small memory requirement and 6-fold speedup compared to mere CPU. Parallelization of this algorithm to multiple GPUs could speed the task significantly more. Another new fast small-memory algorithm of this task is that of Khan *et al.* (2009), leveraging sparse suffix arrays for much reduced memory requirement. This algorithm is a good candidate for GPU implementation and could significantly speed up the MumMer genome alignment algorithm. Exactly matching subsequences are very useful anchoring points in genome to genome alignments as seeds for further optimization of gapped alignments in both directions from the exactly matched segment. In fact, very recently the first such MumMer implementation (MumMerGPU v. 2.0) was released (Trapnell & Schatz 2009) and showed three-fold acceleration compared to CPU based suffix tree methods. Further improvement is to be expected.

Further examples of GPU algorithms for biodata include distributed genetic algorithms (Harding & Banzhaf 2009) and GPU based analysis of mass spectrometric proteome data by adaptive wavelet transforms (Hussong *et al.* 2009), as well as RNA-RNA interaction computation by GPU acceleration to find a stable conformation of two RNAs binding to each other. A 2.5-fold performance boost was reached by GPU implementation compared to a similar cost CPU based system (Rizk & Lavenier 2008).

Finally, computational biochemistry is already benefiting from GPU algorithms. For example, protein solvent accessible surface and desolvation has been computed by GPUs (Dynerman *et al.* 2009) and molecular dynamics simulations of large proteins can be accelerated by a GPU implementation. Speed-up of up to 700-fold compared to a single CPU was found by Friedrichs *et al.* (2009). Similarly, Harvey *et al.* (2009) could simulate large proteins with tens of thousands of atoms for biologically relevant timescales (tens of nanoseconds) in one day on a standard GPU. Such methods show great promise in general.

It is noteworthy, that even these highly accelerated algorithms still use only a small fraction of the potential throughput of the gigaflops and teraflops of processing cycles available in modern

GPU boards. It may be that also in bioinformatics data processing we are approaching the so called memory wall, where the speed of memory (data flow to the processor) is becoming a limiting factor, rather than the clock speed. Current memory speeds are about 100 nanoseconds, while CPU clock cycles are mere nanoseconds, almost two orders of magnitude lower.

One way to avoid the memory wall problem is to parallelize the GPUs. Distributed CPU grid computing is actually moving to this direction, *e.g.* GPUGRID at www.gpugrid.net can already deliver over 30 teraflops of sustained speed for various biological and physical algorithms. The next step then may be to move to cloud computing (cloud computing = scheduled virtual machines running in a grid) using distributed GPU resources or a mixture of heterogeneous CPU/GPU/storage facilities.

Virtual machines thus seem poised to become another useful platform technology for bioinformaticians, who want to analyse huge datasets by cloud computing without buying their own expensive hardware for short-term projects. There are several portable Linux operating systems for bioinformatics that can be run from CD-ROM or USB memory stick, and some of them can already be run as virtual machines (BioSLAX and DNALinux, see Rana & Foscarini 2009). This way one could replicate the virtual machine on remote machines for virtualized cloud computing without one's own hardware. Such a system has already been set up using SnowFlock virtual machine cloning and MPI parallelization (Patchin *et al.* 2009) and using XMPP (Wagener *et al.* 2009). Langmead *et al.* (2009b) already used such a system for single nucleotide polymorphisms in large-scale sequencing data.

In summary, the forthcoming age of the GPU machines and virtual machine based cloud computing will enable bioinformaticians to tackle much more demanding and ambitious computational challenges, because much larger problems can be solved by even brute force. This includes problems that can be transformed into non-negative matrix transformations (NMF, see subchapter 6.6.3.2 for details) complete enumeration of all possible solutions or deep sampling of large data spaces by *e.g.*

genetic algorithms or quantum computing inspired methods to find optimal solutions. It seems that in the next decade of high-throughput bioinformatics virtualization and GPU algorithms will play an important role, at least in the industrial strength application of data analysis for large-scale genomic diagnostics, personalized medicine and metagenomics projects. Thus at least in these subfields of biodata analysis, bioinformaticians will be challenged to learn from and liaise with the high performance computing experts in grid and cloud computing disciplines.

References

Abderrahim, A., Talbi, E-G. & Khaled, M. 2009. Hybridization of Genetic and Quantum Algorithm for gene selection and classification of Microarray data. *Proceedings of IPDPS, pp.1–8, 2009 IEEE International Symposium on Parallel & Distributed Processing.*

Adjeroh, D., Bell, T. & Mukherjee, A. 2008. The Burrows–Wheeler Transform: Data Compression, Suffix Arrays, and Pattern Matching. *Springer, 352 pp.*

Aji, A.M. & Feng, W-C. 2008a. Optimizing performance, cost, and sensitivity in pairwise sequence search on a cluster of PlayStations. *8th IEEE International Conference on BioInformatics and BioEngineering, Athens, pp. 1–6. ISBN: 978-1-4244-2844-1*

Aji, A.M. & Feng, W-C. 2008b. Accelerating Data-Serial Applications on Data-Parallel GPGPUs: A Systems Approach. *Technical Report TR-08-24, Computer Science, Virginia Tech.*

Alexopoulou, D., Andreopoulos, B., Dietze, H., Doms, A., Gandon, F., Hakenberg, J., Khelif, K., Schroeder, M. & Wächter, T. 2009. Biomedical word sense disambiguation with ontologies and metadata: automation meets accuracy. *BMC Bioinformatics 10: 28.*

Alves, R., Vilaprinyo, E. & Sorribas, A. 2008. Integrating Bioinformatics and Computational Biology: Perspectives and Possibilities for In Silico Network Reconstruction in Molecular Systems Biology. *Current Bioinformatics 3(2): 98–129.*

Andreopoulos, B., An, A., Wang, X. & Schroeder, M. 2009. A roadmap of clustering algorithms: finding a match for a biomedical application. *Brief Bioinform. 10(3): 297–314.*

Anon. 2009. http://www.scalableinformatics.com/news/introducing-pegasus-gpu

Armañanzas, R., Inza, I., Santana, R., Saeys, Y., Flores, J.L., Lozano. J.A., Peer, Y.V., Blanco, R., Robles, V., Bielza, C. & Larrañaga, P. A review of estimation of distribution algorithms in bioinformatics. *BioData Min. 1(1): 6.*

Arnau, V., Gallach, M., Marín, I. 2008. Fast comparison of DNA sequences by oligonucleotide profiling. *BMC Res Notes. 1: 5.*

Banzhaf, W., Harding, S., Langdon, W.B. & Wilson, G. 2009. Accelerating Genetic Programming through Graphics Processing Units. *In: Genetic and Evolutionary Computation, Genetic Programming Theory and Practice VI, Springer, pp. 1–19.*

Bayati, M., Gerritsen, M., Gleich, D., Saberi, A. & Wang, Y. 2009. Algorithms for Large, Sparse Network Alignment. *Proceedings of IEEE International Conference on Data Mining (ICDM) (to appear).*

Bloom, J.S., Khan, Z., Kruglyak, L., Singh, M. & Caudy, A.A. 2009. Measuring differential gene expression by short read sequencing: quantitative comparison to 2-channel gene expression microarrays. *BMC Genomics. 10(1): 221.*

Buehrer, G., Parthasarathy, S. & Goyder, M. 2008. Data mining on the cell broadband engine. *Proceedings of the 22nd annual international conference on Supercomputing, Island of Kos, Greece, pp. 26–35.*

Chang, D., Jones, N.A., Li, D., Ouyang, M. & Ragade, R.K. 2008. Compute Pairwise Euclidean Distances of Data Points with GPUs. *Proceedings of the Symposium for Intelligent Systems and Control (ISC 2008), p. 634–017.*

Chen, H., Mao, Y., Zheng, X., Cui, M., Feng, Y., Deng, S., Yin, A., Zhou, C., Tang, J., Jiang, X. & Wu Z. 2007. Towards Semantic e-Science for Traditional Chinese Medicine. *BMC Bioinformatics 8 Suppl 3: S6.*

Chen, C., Schmidt, B., Weiguo, L. & Müller-Wittig, W. 2008. GPU-MEME: Using Graphics Hardware to Accelerate Motif Finding in DNA Sequences. *Lecture Notes in Computer Science 5265: 448–459.*

Cheung, K.H., Frost, H.R., Marshall, M.S., Prud'hommeaux, E., Samwald, M., Zhao, J. & Paschke, A. 2009. A journey to Semantic Web query federation in the life sciences. *BMC Bioinformatics 10 Suppl 10: S10.*

Chong, J., Yi, Y., Faria, A., Satish, N.R. & Keutzer, K. 2008. Data-Parallel Large Vocabulary Continuous Speech Recognition on Graphics Processors. *Technical Report No. UCB/EECS-2008-69* http://www.eecs.berkeley.edu/Pubs/TechRpts/ 2008/EECS-2008-69.html

Dess, N. & Pes, B. 2009. An Evolutionary Method for Combining Different Feature Selection Criteria in Microarray Data Classification. *Journal of Artificial Evolution and Applications 2009, Article ID 803973.*

Dietze, H., Schroeder, M. 2009. GoWeb: a semantic search engine for the life science web. *BMC Bioinformatics 10 Suppl 10: S7.*

Duval, B. & Hao, J.K. 2009. Advances in metaheuristics for gene selection and classification of microarray data. *Brief Bioinform. 2009 Sep 29. [Epub ahead of print]*

Dynerman, D., Butzlaff, E. & Mitchell, J.C. 2009. CUSA and CUDE: GPU-Accelerated Methods for Estimating Solvent Accessible Surface Area and Desolvation. *Journal of Computational Biology 16(4): 523–537.*

Ferragina, P., González, R., Navarro, G. & Venturini, R. 2009a. Compressed text indexes: From theory to practice *Journal of Experimental Algorithmics (JEA) 13(12).*

Ferragina, P., Gagie, T. & Manzini, G. 2009b. Lighweight data indexing and compression in external memory. *arXiv:0909.4341*

Friedrichs, M.S., Eastman, P., Vaidyanathan, V., Houston, M., Legrand, S., Beberg, A.L., Ensign, D.L., Bruns, C.M. & Pande, V.S. 2009. Accelerating molecular dynamic simulation on graphics processing units. *J Comput Chem. 30(6): 864–72.*

Gnirke, A., Melnikov, A., Maguire, J., Rogov, P., LeProust, E.M., Brockman, W., Fennell, T., Giannoukos, G., Fisher, S., Russ, C., Gabriel, S., Jaffe, D.B., Lander, E.S. & Nusbaum, C. 2009. Solution hybrid selection with ultra-long oligonucleotides for massively parallel targeted sequencing. *Nature Biotechnol. 27(2): 182–189.*

Gotoh, O. 2008a. A space-efficient and accurate method for mapping and aligning cDNA sequences onto genomic sequence. *Nucleic Acids Res. 36(8): 2630–2638.*

Gotoh, O. 2008b. Direct mapping and alignment of protein sequences onto genomic sequence. *Bioinformatics.24(21): 2438–2444.*

Harding, S.L. & Banzhaf, W. 2009. Distributed genetic programming on GPUs using CUDA. *Genetic Programming and Evolvable Machines (to appear).*

Harvey, M.J., Giupponi, G.G. & De Fabritiis, G. 2009. ACEMD: Accelerating bio-molecular dynamics in the microsecond time-scale. *arXiv:0902.0827v1 [physics.comp-ph]*

Hastie, T., Tibshirani, R. & Friedman, J. 2009. The elements of statistical learning theory. *New York, NY: Springer.*

Havukkala, I. 2009. Ontologies and semantic mining for bio-technology and chemistry data and patents. *Proceedings of the 2nd international workshop on Patent information retrieval, Hong Kong, China, pp. 41–42.*

Hussong, R., Gregorius, B., Tholey, A. & Hildebrandt, A. 2009. Highly accelerated feature detection in proteomics data sets using modern graphics processing units. *Bioinformatics. 2009 May 14. [Epub ahead of print]*

Jang, H., Park, A. & Jung, K. 2008. Neural Network Implementation using CUDA and OpenMP. *Proceedings of the 2008 Digital Image Computing: Techniques and Applications – pp. 155–161.*

Jégou, H., Douze, M. & Schmid, C. 2009. Searching with quantization: approximate nearest neighbor search using short codes and distance estimators. *INRIA report N0° 7020.*

Jung, S. 2009. Parallelized pairwise sequence alignment using CUDA on multiple GPUs. *BMC Bioinformatics 10(Suppl 7): A3.*

Karanam, R.K., Ravindran, A. & Mukherjee, A. 2008. A stream chip-multiprocessor for bioinformatics. *ACM SIGARCH Computer Architecture News 36(2): 2–9.*

Kasabov, N., Song, Q., Benuskova, L., Gottgtroy, P.C.M., Jain, V., Verma, A., Havukkala, I., Rush, E., Pears, R., Tjahjana, A., Hu, Y. & MacDonell, S. 2008. Integrating Local and Personalised Modelling with Global Ontology Knowledge Bases for Biomedical and Bioinformatics Decision Support. *Computational Intelligence in Biomedicine and Bioinformatics 2008: 93–116.*

Khan, Z., Bloom, J.S., Kruglyak, L. & Singh, M. 2009. A Practical Algorithm for Finding Maximal Exact Matches in Large Sequence Data Sets Using Sparse Suffix Arrays. *Bioinformatics. 2009 Apr 23. [Epub ahead of print]*

Kloetzli, J., Strege, B., Decker, J. & Olano, M. 2008. Parallel Longest Common Subsequence using Graphics Hardware. *In: J. Favre, K. - L. Ma, and D. Weiskopf (Editors), Proceedings, Eurographics Symposium on Parallel Graphics and Visualization, April 14–18, Crete, Greece.*

Kurtz, S., Narechania, A., Stein, J.C. & Ware, D. 2008. A new method to compute K-mer frequencies and its application to annotate large repetitive plant genomes. *BMC Genomics. 9: 517.*

Lancashire, L.J., Lemetre, C. & Ball, G.R. 2009. An introduction to artificial neural networks in bioinformatics--application to complex microarray and mass spectrometry datasets in cancer studies. *Brief Bioinform. 10(3): 315–329.*

Lanckriet, G.R., De Bie, T., Cristianini, N., Jordan, M.I. & Noble, W.S. 2004. A statistical framework for genomic data fusion. *Bioinformatics 20(16): 2626–2635.*

Langdon, W.B. & Harrison, A.P. 2008. GP on SPMD parallel graphics hardware for mega Bioinformatics data mining *Soft Computing 12(12): 1169–1183.*

Langdon, W.B. 2008. Evolving GeneChip Correlation Predictors on Parallel Graphics Hardware. *Proceedings of IEEE CEC 2008, Hong Kong, pp. 4152–4157.*

Langmead, B., Trapnell, C., Pop, M. & Salzberg, S. 2009a. Ultrafast and memory-efficient alignment of short DNA sequences to the human genome. *Genome Biology 10(3): R25.*

Langmead, B., Schatz, M.C., Lin, J., Pop, M. & Salzberg, S.L. 2009b. Searching for SNPs with cloud computing. *Genome Biol. 10(11): R134.*

Lemetre, C., Lancashire, L.J., Rees, R.C. & Ball, C.R. 2009. Artificial Neural Network Based Algorithm for Biomolecular Interactions Modeling. *Lecture Notes in Computer Science 5517: 877–885.*

Li, R., Li, Y., Kristiansen, K. & Wang, J. 2008a. SOAP: short oligonucleotide alignment program. *Bioinformatics 24: 713–714.*

Li, H., Ruan, J. & Durbin, R. 2008b. Mapping short DNA sequencing reads and calling variants using mapping quality scores. *Genome Res 18: 1851–1858.*

Liang, Y. & Kelemen, A. 2009. Time lagged recurrent neural network for temporal gene expression classification. *Int. J. Comput. Intelligence in Bioinformatics and Systems Biology. 1(1): 86–99.*

Liu, Y.I., Wise, P.H. & Butte, A.J. 2009a. The "Etiome": identification and clustering of human disease etiological factors. *BMC Bioinformatics 10(Suppl 2): S14.*

Liu, Y., Maskell, D.L. & Schmidt, B. 2009b. CUDASW++: optimizing Smith-Waterman sequence database searches for CUDA-enabled graphics processing units. *BMC Res Notes 2: 73.*

Ma, W. & Agrawal, G. 2009. A translation system for enabling data mining applications on GPUs. *Proceedings of the 23rd international conference on Supercomputing, pp. 400–409.*

Mäkinen, V., Navarro, G., Siren, J. & Välimäki, N. 2009a. Storage and Retrieval of Individual Genomes. *Journal of Computational Biology (in press).*

Mäkinen, V., Navarro, G., Sirén, J. & Välimäki, N. 2009b. Storage and Retrieval of Highly Repetitive Sequence Collections. *Journal of Computational Biology (in press).*

Manavski, S.A. & Valle, G. 2008. CUDA Compatible GPU Cards as Efficient Hardware Accelerators for Smith-Waterman Sequence Alignment. *BMC Bioinformatics 9(Suppl 2): S10.*

Manavski, S.A., Albiero, A., Forcato, C., Vitulo, N. & Valle, G. 2008. GEAgpu: Improved alignment of spliced DNA sequences to genomic data using Graphics Processing Units. *ECCB 2008, September 22nd–26th, Sardinia – Italy*

Mons, B. & Velterop, J. 2009. Nano-Publication in the e-science era. *Workshop SWAT4LS Semantic Web Applications and Tools for Life Sciences, Proceedings (in press).*

Motsinger-Reif, A.A. & Ritchie, M.D. 2008. Neural networks for genetic epidemiology: past, present, and future. *BioData Min. 1(1): 3.*

Muselli, M., Costacurta, M. & Ruffino, F. 2009. Evaluating switching neural networks through artificial and real gene expression data. *Artif Intell Med. 45(2–3): 163–171.*

Nadler, B., Lafon, S., Coifman, R.R. & Kevrekidis, I.G. 2005. Diffusion maps, spectral clustering and eigenfunctions of Fokker-Planck operators. *In: Advances in neural information processing systems, (eds Y. Weiss, B. Schölkopf & J. Platt), 18: 955–962.*

Nguyen, V.H. & Lavenier, D. 2008. Speeding up subset seed algorithm for intensive protein sequence comparison. *IEEE International Conference on Research, Innovation and Vision for the Future, RIVF 2008 pp. 57–63.*

Nicolas, F., Mäkinen, V. & Ukkonen, E. 2009. Efficient construction of maximal and minimal representations of motifs of a string. *Theor. Comput. Sci. 410(30–32): 2999–3005.*

Noy, N.F., Shah, N.H., Whetzel, P.L., Dai, B., Dorf, M., Griffith, N., Jonquet, C., Rubin, D.L., Storey, M.A., Chute, C.G. & Musen, M.A. 2009. BioPortal: ontologies and integrated data resources at the click of a mouse. *Nucleic Acids Res. 37(Web Server issue): W170–173.*

Oehmen, C.S. & Nieplocha, J. 2006. ScalaBLAST: A scaleable implementation of BLAST for high-performance data-intensive bioinformatics analysis. *IEEE Transactions on Parallel and Distributed Systems. 17: 740–749.*

Onuki, R., Shibuya, T. & Kanehisa, M. 2009. New Kernel Methods for Phenotype Prediction from Genotype Data. *Genome Informatics 22: 132–141.*

Orriols-Puig, A., Casillas, J. & Bernadó-Mansilla, E. 2008. Genetic-based machine learning systems are competitive for pattern recognition. *Evolutionary Intelligence 1(3): 209–232.*

Patchin, P., Lagar-Cavilla, H.A., de Lara, E. & Brudno, M. 2009. Adding the easy button to the cloud with SnowFlock and MPI. *European Conference on Computer Systems, Proceedings of the 3rd ACM Workshop on System-level Virtualization for High Performance Computing, pp. 1–8.*

Platel, M.D., Schliebs, S. & Kasabov, N. 2010. Quantum-Inspired Evolutionary Algorithm: A Multimodel EDA. *IEEE Transactions on Evolutionary Computation 26: 215–145. E-reprint 2009.*

Rana, A. & Foscarini, F. 2009. Linux distributions for bioinformatics: an update. *EMBnet.news 15.3: 35–41.*

Reeves, G.A., Talavera, D. & Thornton, J.M. 2009. Genome and proteome annotation: organization, interpretation and integration. *J. R. Soc. Interface 6(31): 129–147.*

Rizk, G. & Lavenier, D. 2008. GPU Accelerated RNA-RNA Interaction Algorithm. *EMBnet Conference 2008 - 20th Anniversary Celebration, Abstract 7.*

Samwald, M. & Stenzhorn, H. 2009. Simple, ontology-based representation of biomedical statements through fine-granular entity tagging and new web standards. *Bio-Ontologies 2009 conference (in press).*

Samwald, M., Lim, E., Masiar, P., Marenco, L., Chen, H., Morse, T., Mutalik, P., Shepherd, G., Miller, P. & Cheung, K-H. 2009. Entrez Neuron RDFa: a pragmatic Semantic Web application for data integration in neuroscience research. *Studies in Health Technology and Informatics 150: 317–321.*

Schmidt, B., Sinha, R., Beresford-Smith, B. & Puglisi, S. 2009. A Fast Hybrid Short Read Fragment Assembly Algorithm. *Bioinformatics (in press).*

Schütte, M., Klitgord, N., Segrè, D. & Ebenhöh, O. 2009. Co-Evolution of Metabolism and Protein Sequences. *Genome Informatics 22: 156–166.*

Shah, A.R., Markowitz, V.M. & Oehmen, C.S. 2007. High-throughput computation of pairwise sequence similarities for multiple genome comparisons using ScalaBLAST. *Life Science Systems and Applications Workshop, 2007. LISA 2007. IEEE/NIH, pp. 89–91.*

Singh, R. *et al.* 2008. Global alignment of multiple protein interaction networks with application to functional orthology detection. *Proc. Natl Acad. Sci. USA, 105, 12763–12768.*

Siren, J. 2009. Compressed Suffix Arrays for Massive Data. Lecture Notes in Computer Science 5271: 63–74.

Skillikorn, D. 2005. Understanding complex datasets: Data mining with matrix decompositions. *Chapman & Hall.*

Striemer, G.M. & Akoglu, A. 2009. Sequence alignment with GPU: Performance and design challenges. *pp. 1–10, 2009 IEEE International Symposium on Parallel & Distributed Processing.*

Sutherland, K., McLeod, K., Ferguson, G. & Burger, A. 2009. Knowledge-driven enhancements for task composition in bioinformatics. *BMC Bioinformatics 10 Suppl 10: S12.*

Szalkowski, A., Ledergerber, C. Krähenbühl, C.P. & Dessimoz, C. 2008. SWPS3 - fast multi-threaded vectorized Smith-Waterman for IBM Cell/B.E. and ×86/SSE2, *BMC Research Notes 1: 107.*

Tamura, T., Christian, N., Takemoto, K., Ebenhöh, O. & Akutsu, T. 2009. Analysis and Prediction of Nutritional Requirements Using Structural Properties of Metabolic Networks and Support Vector Machines. *Genome Informatics 22: 176–190.*

Trapnell, C. & Schatz, M.C. 2009. Optimizing data intensive GPGPU computations for DNA sequence alignment *Parallel Computing 35(8–9): 429–440.*

Trapnell, C., Pachter, L. & Salzberg, S.L. 2009. TopHat: discovering splice junctions with RNA-Seq. *Bioinformatics. (e-print, March 16 2009).*

Tsafrir, D., Tsafrir, I., Ein-Dor, L., Zuk, O., Notterman, D.A. & Domany, E. 2005. Sorting points into neighborhoods (SPIN): data analysis and visualization by ordering distance matrices. *Bioinformatics 21(10): 2301–2308.*

Tsutsui, S. & Fujimoto, N. 2009. Solving quadratic assignment problems by genetic algorithms with GPU computation: a case study. *Proceedings of the 11th Annual Conference Companion on Genetic and Evolutionary Computation Conference, 2523–2530.*

von Luxburg, U., Belkin, M. & Bousquet, O. 2008. Consistency of spectral clustering. *Ann. Statist. 36(2): 555–586.*

Wagener, J., Spjuth, O., Willighagen, E.L. & Wikberg, J.E. 2009. XMPP for cloud computing in bioinformatics supporting discovery and invocation of asynchronous web services. *BMC Bioinformatics 10: 279.*

Wirawan, A., Kwoh, C.K., Hieu, N.T. & Schmidt, B. 2008. CBESW: Sequence Alignment on the Playstation 3. *BMC Bioinformatics 9: 377.*

Yao, L. & Rzhetsky, A. 2008 Quantitative systems-level determinants of human genes targeted by successful drugs. *Genome Res. 18(2): 206–213.*

Yu, S., De Moor, B. & Moreau, Y. 2009. Clustering by heterogeneous data fusion: framework and applications. *In: Neural Information Processing (NIPS workshop, Whistler, Canada, Dec. 2008).*

Xiao, J., Xu, J., Chen, Z., Zhang, K. & Pan, L. 2009. A hybrid quantum chaotic swarm evolutionary algorithm for DNA encoding. *Computers & Mathematics with Applications 57(11–12): 1949–1958.*

Zhang, H., Schmidt, B. & Müller-Wittig, W. 2008. Accelerating BLASTP on the Cell Broadband Engine. *In: M. Chetty, A. Ngom, and S. Ahmad (Eds.): PRIB 2008, LNBI 5265, pp. 460–470.*

Zhao, J., Miles, A., Klyne, G. & Shotton, D. 2009a. Linked data and provenance in biological data webs. *Brief Bioinform 10(2): 139–152.*

Zhao, J., Jentzsch, A., Samwald, M. & Cheung, K-H. 2009b. Linked Data for Connecting Traditional Chinese Medicine and Western Medicine. *The Sixth International Workshop of Data Integration in the Life Sciences, Manchester, UK, 2009 (poster).*

Index